CHICAGO PUBLIC LIBRARY
BEVERLY BRANCH
1962 W. 95th STREET
CHICAGO, IL   60643

DEC   1998

DISCARD

CHICAGO PUBLIC LIBRARY
BEVERLY BRANCH
1962 W. 95th STREET
CHICAGO, IL   60643

IRISH STONE WALLS

CHICAGO PUBLIC LIBRARY
BEVERLY BRANCH
1962 W. 95th STREET
CHICAGO, IL   60643

Patrick McAfee is an expert stonemason. Born and brought up in Dublin where he served an apprenticeship to his father, he worked for a number of years in Australia and studied traditional methods of working with stone and lime mortars at the European School of Conservation at San Servolo in Venice. He now divides his time between workshops on stone and lime and consultancy work around the country. He hopes that *Irish Stone Walls* will be a useful guide to amateur and professional stone workers alike and an accessible history on the use of stone in Ireland throughout the centuries.

RO7125 95165

IRISH STONE WALLS

History ▪ Building ▪ Conservation

Patrick McAfee

CHICAGO PUBLIC LIBRARY
BEVERLY BRANCH
1962 W. 95th STREET
CHICAGO, IL   60643

THE O'BRIEN PRESS
DUBLIN

First published 1997 by The O'Brien Press Ltd,
20 Victoria Road, Rathgar, Dublin 6, Ireland.

Published in US and Canada by
The Irish American Book Company
Monarch Park Place
Niwot, Colorado 80503
1800 352 1985

Copyright © for text, photographs and illustrations – Patrick McAfee
Copyright © for layout, design and editing – The O'Brien Press Ltd

1 2 3 4 5 6 7 8 9 10
97 98 99 00 01 02 03 04 05 06

All rights reserved. No part of this book may be reproduced
or utilised in any form or by any means, electronic or mechanical,
including photocopying, recording or by any information storage and
retrieval system without permission in writing from the publisher.

British Library Cataloguing-in-Publication Data
McAfee, Pat
Irish stone walls: history, building, restoring
1.Stone walls – Ireland
I.Title
721.2'09415

ISBN 0-86278-478-6

The O'Brien Press receives assistance from
The Arts Council/An Chomhairle Ealaíon

This publication is sponsored by the Heritage Council

Editing, layout, design: The O'Brien Press Ltd
Cover photograph: The Department of Arts, Culture and the Gaeltacht
Cover separations: Lithoset Ltd
Printing: Hartnolls Ltd

**This book is dedicated to
Ann, Brian and Barbara**

Acknowledgements

I would like to thank my family – Ann, Brian and Barbara for their support and for putting up with me while I was writing this book. To Greta and Hugh, my parents, and to the early days when I was apprenticed to my father and learned so much.

To members of FÁS – Eamon Rapple, Jane Forman, Pat Kelly Rodgers, Fred English, Assunta Kavanagh, Noel Feeney, Mary Liddy, Jim Casey and Gerald Rynne who showed continous support and belief in me over the years.

To members of Cornerstone – Conor Rush, Claire Gogarty, Anne Lodge and David Johnson for working together with such cooperation to achieve a dream.

To Peter Pearson for his love of old buildings.

To Peter Geraghty, formerly of the Office of Public Works, for his technical knowledge of medieval buildings.

To Amanda Wilton for the space and support while writing the first draft of this book. To Jim Farrell and Joe Carroll.

To Aubrey Flegg of the Irish Geology Survey (IGS) for his enthusiasm about geology.

To Nioclás Breathnach who taught me everything I know about *Bearlager na Saor*.

To Bairbre ní Fhloinn and Criostóir mac Carthaigh in the Department of Irish Folklore at University College Dublin for their assistance.

To Mairéad Ashe-FitzGerald for advice on the history of stone.

And last but not least to the directors and staff especially Lynn Pierce and Frances Power at The O'Brien Press.

To the Heritage Council for their support.

CONTENTS

Foreword **9**

Introduction **10**

1. The History of Stone in Ireland **13**
2. Geology **37**
3. Rubble Walling Styles **42**
4. Tools **48**
5. Selecting Rubble Stone from the Quarry **52**
6. Transporting Stone **57**
7. Site Organisation **59**
8. Bedding Stone **61**
9. Types of Profile **63**
10. Plumb Bobs **66**
11. Foundations **70**
12. Bonding **74**
13. Copings **80**
14. Cutting Rubble Stone **84**
15. Cutting Quoin Stones **88**
16. Surface Finishes **94**
17. Dry Stone Walling **99**
18. Retaining Walls **109**
19. Circular Piers **118**
20. Lime Mortars **123**
21. Pointing **136**
22. Wet Dashing **150**
23. Structural Problems **154**
24. Bad Habits in Modern Stone Walling **157**
25. Inappropriate Modern Repairs **159**
26. Quantities, Weights and Measures **161**

Suppliers and Services **163**

Glossary **164**

Bibliography **169**

Index **172**

Oratory of Gallarus, Dingle Peninsula, County Kerry: sandstone laid dry in the Early Christian period.

FOREWORD

Whether in rugged dry stone walls, whitewashed farm buildings or the smooth rendered walls of streetscapes, stone is the indigenous building material of the Irish town and countryside. It has a history stretching back in time. The crafts of stone building and lime rendering were timeless skills, built up by years of experience and handed on from generation to generation.

The arrival of modern building materials over the last fifty years and the lingering association of stone cottages with poverty have meant that the tradition of building in stone has been all but lost to this generation.

Pat McAfee has recognised the good sense and beauty of building in stone and rendering in lime, and has quietly worked on reviving the tradition while many others turned their backs on it. He has learned, taught, counselled, advised, and studied traditional building methods, and this book is the result of years of patience as enduring as the stone itself. His book tells the history of stone walls in Ireland and describes the traditions that grew up in different regions. He describes with clarity and authority the construction of stone walls from foundation to coping stones – quarrying and selection of stone, the tools for dressing and pointing stone, lime mortar mixes, pointing and rendering techniques.

Anyone who is involved – either through an historical interest, or because of a responsibility for the care and maintenance of our built heritage – will find in this book the answers to the practical questions, gain an understanding of our built heritage and discover a whole new way of 'reading' the buildings we so often take for granted. Architects, builders, local authorities, community groups, students of folklore, history and Irish traditions will find and, I hope, pass on the almost lost traditions of Irish building methods.

The Heritage Council is proud to sponsor this publication, recognising the growing awareness and hunger for practical information through the hundreds of requests it receives.

In a period of building 'boom', this book is a timely reminder to us to retain and respect the legacy we have been given and enable us to hand it on to future generations.

Freda Rountree
Chairperson of the Heritage Council

An Chomhairle Oidhreachta   The Heritage Council

INTRODUCTION

'Art grows out of good work done by men who enjoy it. It is the wealth, surely, of any country.'

From *Stone Mad* by Seamus Murphy

Stone is a material inextricably linked to the history of the human race. It has offered shelter from the climate, protection from enemies, housed gods, reflected the wealth of kings and recorded the coming and going of the seasons. It has been used to make both tools and weapons of war, to make the wheel to grind corn, to carve and sculpt and to commemorate the honoured and the dead. The act of accomplishing anything of size in stone involves the organisation of labour and other support services. A civilisation that achieved high standards of work in stone therefore was invariably one that was relatively stable and sophisticated.

In Ireland there is a history of working with stone that stretches back at least 5,000 years to some of the finest dry built structures in the country, such as Newgrange in the Boyne Valley complex where the spirals carved in stone are as clear today as they were when first done – in places the pecking pattern from the sharp stone tools looks like the workers have just left. It often comes as a surprise then that many of the thousands of miles – about 250,000 miles altogether – of grey stone walls which, set against a backdrop of heather-covered mountain, green fields and grey skies, haunt the memory of both visitors and Irish emigrants alike are no more than 150 years old, and were built after the break-up of the Rundale Village system of open farming.

Whether ancient or relatively modern, this simple and anonymous stone work needs to be preserved and repaired sympathetically when necessary, because it acts as a link to the past, holding a key to Irish history and culture. Some of the walls are unexpected works of art not just because of the cutting or dressing that went into them, but for the care and sympathy that was put into their construction, the respect shown to the material itself, and the relationship of this material with the land.

Many communities are now embarking on the repair of their old town and village walls and on building new walls, while miles of stone walls are being built alongside motorways. Individuals are also realising the durability, beauty and value of stone and are using it to construct walls around their houses, or building stone entrances.

Irish Stone Walls is about building new walls in the traditional ways and repairing old walls sympathetically, using the appropriate materials and methods. When it comes to repair, in particular, there is a need to examine what was done in the past more closely rather than to impose present-day interpretations of what a stone wall should look like. This is partly due to the fact that traditional stonemasonry methods have largely fallen into disuse or been forgotten, so that modern materials and ways of working are being applied to older work, often with disastrous results.

This book seeks to introduce some basic principles into the building and repair of stone walls. At the moment there is little discrimination being made between what is and is not acceptable. The walls considered here are the boundary walls that separate one property from another or a private property from a public road, the high walls around estates and those at the entrance to towns and villages which were sometimes built during the famine years of 1845-49 to give much-needed work to the local tenantry. The entrances to these estates are usually of a high standard, at least equal to if not surpassing the quality of work on the house itself.

A couple of points: Care must be taken about making changes to some old walls, such as those that surround graveyards, and any wall that has a historic significance, such as town walls which once had a defensive purpose. These come under the control of the Department of Arts, Culture and the Gaeltacht/Heritage as do other monuments and sites listed in the Sites and Monuments Record (SMR). Should any object be found – cut, dressed or carved stone, for example – both the National Museum and the Department of Arts, Culture and the Gaeltacht/Heritage Services must be notified.

The terms 'ditch' and 'fence' are sometimes used when referring to walls. In order to prevent misunderstanding, the terms 'dry stone walls' and 'mortared walls' are used throughout this text. Freehand drawings along with photographs have been used whenever possible as they bridge the gap between the written word and the practice of doing. They have been arranged in a logical sequence so as to make the practice of repair and conservation and building more easily understood.

I hope *Irish Stone Walls* will be useful both on-site and in the office.

Pat McAfee
April 1997

The Mourne Wall, County Down: 35 kilometres of dry stone walling, cut from locally available field stone using plugs and feathers.

CHAPTER 1

THE HISTORY OF STONE IN IRELAND

The ways in which stone was worked and used and developed as a technology reflect the story of our ancestors in prehistoric times.

THE AXE MAKERS

No stone houses or human remains have been found of the first people to come to Ireland, the Mesolithic or Middle Stone Age people, in about 7000BC. But their stone artefacts provide us with useful information about their hunting and fishing lifestyles. One of the places where they camped in their skin houses was at Mount Sandal in County Derry. Flint was the most plentiful material in Mesolithic times for the minutely worked microliths which they used as arrow tips, on knives or for scrapers and needle points for working on skins. They also made rough stone axes.

THE TOMB BUILDERS

The people of the Neolithic or the New Stone Age who arrived in Ireland about 4000BC were farmers. They were expert makers of stone axes, much-needed tools for felling the trees that covered the country. They made polished stone axes out of the porcellanite which was in plentiful supply in the northeast of Ireland. Two sites, one at Tievebulliagh in County Antrim and the other on Rathlin Island off the coast of Antrim were the sources of a flourishing trade in stone axes all over Ireland and Britain.

The Neolithic people lived in farming communities in houses that were rectangular wooden structures. All their stone-building skills – which were advanced – went into the building and decoration of tombs for the dead such as those in the Boyne Valley. Newgrange in County Meath, built about 5,000 years ago, is one of the finest examples in the country (see below). The whole structure is indicative of there being a well thought-out design, organised and skilled labour and a period of relative social stability.

Newgrange, County Meath: a 5,000 year old tomb built without metal tools.

NEWGRANGE

Location: Newgrange, County Meath.

Age: *c* 3,000BC.

Geology: *Quartz* – white in colour, used as a non-structural external facing, probably transported from the Dublin/Wicklow mountains by sea and the river Boyne.

Granite – grey/brown in colour, round/oval in shape and projecting from the quartz facing, possibly transported from as far away as the Mourne mountains in County Down.

Greywacke – green in colour, used in kerbs, corbels and other elements, some with carvings – the largest kerbstone is 4.5m long. This stone was found locally but not quarried as it was weathered before it was placed.

Gravel – and small stones from a local source used as fill which makes up the bulk of the monument.

Points of Interest: Newgrange is a passage tomb built by the

Newgrange: external wall showing quartz, granite and greywacke.

Neolithic or New Stone Age people, which, on the shortest day of the year, 21 December, is aligned to the rising sun. The sun's rays enter the tomb through a light box (or gap) over the doorway and creep up the inclined passage to light the 6m high, inner chamber which is roofed over using stone corbelling. The corbels are inclined outwards and some are chased or grooved on top by approximately 6cm wide and 1cm deep to drain off the rainwater that would otherwise enter the tomb. After 5,000 years the chamber is still dry, a tribute to the skill of its builders. The tomb contains 200,000 tonnes of stone which is faced in white quartz and granite and has a diameter of 79-85m. Some of these stones had to be transported over 80km. Enigmatic spirals and lozenges and other designs are carved onto the kerb stones that surround the outer base perimeter of the tomb.

All carvings were carried out using stone tools, probably made of flint or quartz. The surfaces were 'sparrow picked' using a pointed stone tool that leaves a finish similar to that of a small sharp steel point but, of course, Neolithic man had no metal tools. 'Picking' can be seen not only on the stone carvings but also as a surface finish on the large plain stones along the passage and elsewhere.

Newgrange: enigmatic spirals and lozenges on a kerbstone.

Some of the internal surface finishing was conducted in situ after the stones were positioned.

On excavation the external wall, as shown above, was found to have collapsed. Through trial and error, ie, building a wall of quartz and granite and knocking it down again, a similar collapse pattern was discovered and consequently a reproduction of the original wall (see page 15) could be constructed. How the quartz was originally fixed in position is unknown as no bedding or bonding material was found. This wall also displays a visible batter or inwards inclination.

Newgrange is an unforgettable combination of architecture, engineering and art where the theme of light and shadow not only occurs in the passage and chamber but at more subtle levels in the quartz and projecting granite facing on the southeast elevation and more subtly again in the surface finishes created by sparrow picking. In terms of working with stone this monument has a lot to teach us and many of the skills shown here, such as corbelling and picking, are repeated – though somewhat differently – in subsequent centuries. No real equivalent for the quartz facing occurs until this century.

THE EARLY WALL BUILDERS

The first remnants of stone walls found also date from this period (3000BC). The most famous are those in County Mayo, known as the Céide Fields, but other walls from this period exist under bogs elsewhere in Ireland. In the later Bronze Age period farmers used the ard plough which was pulled by a horse (attached to its tail). Before a field was ploughed the large stones were cleared by hand and used to build walls or stacked in heaps to be cleared later.

CÉIDE FIELDS

Location: The Céide Fields, Ballycastle, County Mayo.
Age: *c* 3000BC.

Céide fields: collapsed and buried stone walls
dating from 5,000 years ago.

Geology: Sandstone.

Points of Interest: Buried stone walled fields set out by Neolithic man about 5,000 years ago and originally used for grazing cattle and growing wheat and barley. This early farming community lived in scattered dwellings much like those of today, clearing the land of trees and building walls to fence off fields.

The overall area of the site is about 10km² – the largest such site in Europe. Some of the walls are 2km long and the total amount of stone is estimated at 250,000 tonnes, although the individual stones are not very large. The way in which the fields are walled into rectangles has not changed significantly. Very few of the walls have been excavated – most remain under the bog and were pinpointed by driving a steel bar down until it struck the stone. Those walls that have been excavated and are now visible show no signs of having been laid to any pattern or bond, and therefore may originally have been stone walls that collapsed over time and were eventually buried under the encroaching bog or even simple heaps of stone that acted as boundary markers after the fields were cleared.

It is possible that because the land had been cleared of trees the earth became increasingly waterlogged and eventually the community could no longer cultivate the area. The wet soil would then have developed into peat bog, eventually covering all trace of both the fields and the walls.

THE FORT BUILDERS

The availability of stone led to the building of the great forts or *dúnta* of the western seaboard. Some of them, at least, had their origins around 1200BC in the Late Bronze Age. The later ring forts were usually built of earth and stone and the remains of over 50,000 are scattered around the country. Their stone counterparts – cashels – are to be seen in hundreds in the western counties especially in the Burren area of County Clare. An archaeological study currently under way on the cashels of the west of Ireland may uncover more about their original purpose.

DÚN AONGHUSA

Location: Dún Aonghusa, Inishmore, Aran Islands, County Galway.
Age: This cliff-edge site was occupied as early as 1300BC. Recent excavations show that the thick inner wall was built around 800BC.
Geology: Carboniferous limestone.

Dún Aonghusa, Inishmore, with its massive fissured dry stone wall in limestone.

Points of Interest: A most distinctive feature of this site is the massive *chevaux de frise*, a ring of enormous, sharp, upright stones that surrounds the fort, providing an effective defence against attack. Buildings in Ireland tended to be circular in shape from prehistoric times up the introduction of Christianity. Circular shapes are naturally more stable, while square structures rely on the structural integrity of their corners which, if lost, leads to collapse, a lesson learned in castle building when miners were used in warfare to undermine castle walls. Forts like Grianán of Aileach, Dún Aonghusa and Staigue Fort are masterpieces of building in stone. They are all circular, with massive battered or inward-inclining walls, thick enough to have guardrooms built into them.

THE EARLY CHRISTIAN PERIOD

With the arrival of Christianity in Ireland in the 5th century, monks such as Saint Kevin who founded Glendalough, turned their backs on society to settle in remote areas. Often they were joined by like-minded pupils and a monastic settlement would grow up with clusters of circular, domed, beehive huts such as still stand on Skellig Michael. The Oratory of Gallarus in the Dingle Peninsula is shaped like an upturned boat and provides a superb example of dry stone work in sandstone.

Small stone churches with steep stone roofs and, often, large face-bedded sedimentary stone, known as **cyclopean**, are characteristic of the period. Extended side walls at the gables known as antae are reminders of the original wooden churches that these stone buildings replaced. A good example is the church on St MacDara's Island off the coast of County Galway. Often finials, again copied from the earlier wooden structures, would be carved as stone crosses.

Up to this period buildings and walls were dry stone, but from now on lime mortars appear, giving greater protection from the rain and wind, and allowing masons to build higher and thinner walls. It is likely that the technique came from the Romans who were masters of lime mortar, and that their influence was now filtering into Ireland. It is possible that the monks who were travelling back and forth to the Continent spreading the Gospel were also returning with new techniques in stone building.

From the 9th to the 12th century round towers such as that at Glendalough were constructed. A noticeable improvement in the standard of work can be seen on some of these towers. Larger individual stones cut on the curve with small joints became common and herald the next period of stone work.

THE CATHEDRAL, GLENDALOUGH

Location: The Cathedral, Glendalough, County Wicklow.
Age: Pre-12th century.

Geology: Mica schist, a green metamorphic stone which was originally a mudstone until, through heat and pressure from the underlying granite, it became metamorphised into harder stone.

The Cathedral, Glendalough, County Wicklow, showing projecting antae and cyclopean stone work in mica schist.

Points of Interest: Cyclopean stone work, as the name suggests, uses large individual stones. This is common practice in Early Christian architecture but is usually associated with the small stone-roofed churches of the period, such as Temple Benan on Inishmore (see Chapter 21: Lime Mortars), and not with large buildings like the cathedral shown above. These stones are face-bedded, displaying their large natural beds (see Chapter 7: Bedding Stone) and giving the impression of being massive stones when in fact they are simply slab-like and bedded on their edges. The church has projecting side walls or antae, like the original wooden churches it mimics. The doorways are narrower in width at the top than the bottom. In general at this period the arches are not true arches made up of individual stones, but are cut from one or two stones only. It is interesting to note that this structure used lime mortar, although there is no local source of limestone with which to produce lime.

THE IRISH ROMANESQUE

In the first half of the 12th century a new style of building in stone, known as Romanesque, appeared in Ireland. At Cormac's Chapel at Cashel in County Tipperary this Romanesque architecture is evident, although it is given a particular Irish style of expression. The true arch – with separate **voussoirs** or individual stones instead of a single or a pair of stones cut into an arch – only appears at this late date. These arches are semi-circular and decorated with carved heads, lozenges, etc, where they occur in prominent positions such as the west doorway to churches and cathedrals. A higher standard of stone work is noticeable with smaller joints and a tendency to cut stones with their **bed** or vertical joints eating into adjacent stones, which indicates that stones are cut to suit their individual position just prior to laying instead of in advance. This is a technique used by the ancient Egyptians and also seen in the work of the Incas in South America. Sandstone was the most popular stone used at this period for cut or decorative details.

The high level of skill displayed in cutting and laying at Clonmacnoise monastery in County Offaly is common from the 12th century when new religious orders such as the Benedictines,

Cistercians and Augustinians and the pilgrims who travelled to France and Italy returned, bringing not only a new organisation to religious life but also the Romanesque style of architecture. Although there was a particular liking for sandstone which was relatively easy to work, granite and limestone were also used. At the 12th century Baltinglass Cistercian abbey all the stonecutting, including quoins, circular and square columns, voussoirs of arches, etc., are cut from a beautifully coloured granite. Ireland is still scattered with many examples of buildings in the Irish Romanesque style such as the church at Kilmalkedar in County Kerry and St Doulagh's Church at Balgriffin in County Dublin. Profusely decorated doorways at the Clonfert Cathedral in County Galway and at the Nun's Chapel at Clonmacnoise are good examples.

CLONFERT CATHEDRAL

Location: County Galway.

Age: 12th century.

Geology: Sandstone.

Points of Interest: The west doorway of Clonfert Cathedral is the supreme example of Irish Romanesque. It displays a rich variety of carving with animal heads, chevrons, zig-zags, palmettes, etc., including human heads or masks.

Clonfert Cathedral, County Galway, the Romanesque west doorway.

CLONMACNOISE ROUND TOWER

Location: Clonmacnoise, County Offaly.

Age: 10th century, rebuilt in 1134.

Geology: Carboniferous limestone.

Points of Interest: A round tower built between the 10th and 12th centuries as a bell tower to summon the monks to prayer and meals. The entrance doors of round towers are high above ground level, implying that the towers may have acted as places of refuge or for storing treasures in times of trouble. The tallest round tower still standing in Ireland is just over 30m high. Like all round towers, the Clonmacnoise one is tapering, with a larger diameter at the base than the top (the average base diameter of these towers is just over 5m). A glance upwards along the outside tapering walls of any of the towers shows just how accurately built most were, with little deviation from a straight line. Even more surprising is their lack of foundations to any depth below ground level.

Clonmacnoise, County Offaly: base of round tower.

It is possible that the method used to build a tapered round tower was to stand a tree trunk which had a natural even taper, upright. From the outer face of the trunk a trammel – a length of wood – could be circumscribed to build a consistently tapering tower. The length of the trammel could be adjusted at various heights for structures that displayed the entasis sometimes seen on classical columns (slight convex curves deliberately used to offset the optical illusion that straight sides tend to give of sloping inwards).

THE ANGLO-NORMANS

In 1169 the Anglo-Normans invaded Ireland and with them came a phase primarily of castle building. Initially the castles were built in timber, sometimes pre-fabricated on a man-made or natural hill called a motte in order to take control of an area quickly. When the invaders became more settled they began building castles in stone. Below the motte was an area, called a bailey, which was surrounded by a ditch, called a fosse if dry, and a moat if it had water. The surrounding land was usually farmed in strips on large open fields by tenants. The first stone castles were well underway before the end of the 12th century – among them were Trim Castle in County Meath and Carrickfergus Castle in County Down. When manor houses were built in later centuries, moats and pallisades were still used for defensive purposes.

In the latter half of the 12th century the Romanesque architecture so popular in ecclesiastical buildings began to make way for the Gothic. Christ Church Cathedral in Dublin provides a perfect example of the transition from one style to another – the north and south transepts are Romanesque while the Gothic is used elsewhere. St Patrick's Cathedral, built only decades later, is entirely Gothic (both were heavily restored in the 19th century). Besides its characteristic pointed arch, the Gothic style also allowed the introduction of more light into a building. New techniques in vaulting transmitted the weight of the roof down clusters of shafts which formed piers or, externally, buttresses (including flying buttresses) and meant that the walls between arches became less structural and could be punctuated more frequently with windows.

The stone used for fine-cutting and decorative details from the 12th to the early 15th century was native sandstone and imported stone, such as Dundry stone which is an oolitic limestone – formed by wave action near a shoreline – from near Bristol in England, or Caen stone from Normandy in France. These soft stones show the marks of being worked with an axe and/or saw. Since stone is heavy it was easier to transport it by ship and barge than by road. Imported oolitic limestone sometimes displays diagonal face dressing, which was produced by an axe or broad chisel and helps in dating the stone. Red sandstone was also used in this period. It was not until the 15th century that Irish carboniferous limestone – a very hard stone – was worked.

The Anglo-Norman masons must have found some of the indigenous stone very hard to work, and more than likely had neither the temper in their tools nor the methodology to tackle them. This is still the case today when stonemasons used to working with soft stone find it very difficult to switch to harder ones.

The 14th century is marked by its absence in Irish architectural history as the Black Death ravaged the country and very little was built.

St Canice's Cathedral, County Kilkenny: the carvings are the work of the Gowran Master.

ST CANICE'S CATHEDRAL

Location: Kilkenny City, County Kilkenny.

Age: Mostly 13th century.

Geology: Part of the west doorway is constructed in oolitic limestone, yellow/cream in colour, imported from Dundry Quarry, 6.4km from Bristol in England.

Points of Interest: The 13th century work of the Gowran Master, a master mason, can be clearly identified here in the carvings and doorways. He carved faces with a similar expressionless look as hood stops (a projecting stone moulding) on all the nave doorways. His work is also to be seen in the tomb in the wall of the north nave and elsewhere in the cathedral. Some of his finest work is to be seen at Gowran in County Kilkenny.

TRIM CASTLE

Location: Trim, County Meath.

Age: 12th century.

Geology: Carboniferous limestone with sandstone reveals.

Points of Interest: Trim Castle was built shortly after the Norman invasion in 1169, and is therefore one of the earliest – and coincidentally the largest – castles in Ireland.

The central keep (tower or donjon) had four projecting smaller towers, one of which is missing. On plan this formed the shape of a Greek cross. Many medieval buildings, including St Canice's Cathedral, were laid out by master masons to an underlying geometry, often with religious symbolism.

Trim Castle, County Meath: the outer curtain wall and barbican.

It is possible that stonecutting for windows and doors for each castle was done elsewhere and then transported to the site. Quoin stones for corners were probably done on-site.

THE 15TH-17TH CENTURIES

Great stone tower houses, a development of the fortified 12th century castle, were built during this period. Rockfleet Castle in County Mayo, home of the legendary Grainne O'Malley, is a fine example of the type. From the early 15th century native Irish limestones were used for doors, windows and architectural detailing and carving, implying that the blacksmith's technology was improving the temper of the stonemason's tools. The mark of the punch is increasingly common on cut stone and over the next two centuries becomes more pronounced and decorative.

In the 16th century drafted margins appeared at corners, windows and doors. Drafted margins or square chiselling are found on the arris (edges) or surfaces of cut stones, revealing a method of achieving a plane surface on a block of stone by 'boning in' out of twist that is still used today (see Chapter 15: Cutting Quoin Stones).

During the Tudor plantations of this period stone-faced banks with ditches were built between fields to prevent cattle rustling and to shelter crops. Ash trees were planted on the banks to provide timber. There was a movement from the countryside to the security of walled towns. Within the Pale – that shifting area around Dublin under English control – open land began to be enclosed and the small plots of land common up until then disappeared. During the Ulster plantation in the early 17th century arable farming became common on small enclosed fields.

ROTHE HOUSE

Location: Rothe House, Kilkenny City, County Kilkenny.
Age: 16th century.
Geology: Carboniferous limestone.
Points of Interest: This is a fine example of a wealthy merchant's house of the period. It shows the distinctive type of decorative punchwork that emerged in the 15th century, first worked in a light

pattern, but increasingly boldly through to the 17th century. A punch about 6mm wide, very similar to that used today, was used to create most of this decorative punchwork. The play of light and shadow on the stone work is very distinctive and attractive. It is possible that some of this tooling was done in order to support an external rendering of lime and aggregate such as a wet dash.

Concentric indented tooling covering the entire surface of Gothic arch stones can also be seen from this period onwards. A native tradition of working Irish limestones for dressings also emerges and limestone is used for all kinds of detailing, including door and window jambs, carved work, etc.

A close-up of decorative punchwork at Rothe House,
County Kilkenny.

The type of decorative punchwork seen here is used to date stone work in Ireland just as the diagonal surface finish of Dundry limestone and red sandstone is used to date stone of the 12th and 13th centuries.

NEWTOWN CASTLE

Location: Ballyvaughan, County Clare.

Age: 16th century.

Geology: Carboniferous limestone.

Points of Interest: A round tower built off a square base. Most Irish castles have a battered base, known as a talus, built to increase the width of the wall and prevent the miners of attacking forces digging underneath the tower to undermine its foundations. A talus also allowed the defenders to drop projectiles such as stones from the tower so that they would bounce out into the faces of the attackers.

Newtown Castle, County Clare: A round tower built off a square base.

At Newtown Castle the square talus makes an unusual angle with the round tower. This defensive element is known as a spur.

THE GEORGIAN ERA

In the early 18th century the Rundale system of open farming flourished, particularly in the western part of the country. Under this system an extended family group living in *clachans* or groups of houses cultivated plots of ground separated by strips of land rather than stone walls. These parcels of land were often spread over a large area, each one being quite small. As the population increased, however, this system became untenable because the land was divided into smaller and smaller plots through inheritance. With the arrival of the agricultural revolution in the mid-18th century landlords hedged their fields, dug ditches, built banks of earth and planted hawthorns to mark field boundaries. From the late 18th century forests and bushes were cut down at an alarming rate for use as fuel. By the time of the famine – except within walled estates – the country looked like a desert.

In the cities, and in particular in Dublin, well-cut ashlar stone with fine joints and classical details was increasingly seen on public buildings, private estates, entrances to estates, etc. Classical proportions and details appeared in every detail, executed in native stones such as limestone, granite and sandstone. Much of the ashlar work is relatively thin, maybe 100-200mm in thickness or less, and backed and bonded into rubble stone or occasionally brick. Many fine Georgian houses were constructed in brick with stone used at the entrances for steps, door surround, etc, all in the classical style. Footpaths and kerbs in Dublin were laid in granite and roads were laid in setts which were imported whinstone (basalt) and Wicklow diorite. Large quantities of slate were also imported for roofing. Portland stone was imported in large quantities from the south of England as drums for columns, window and door surrounds, entablatures, ballusters and many other items. It was often used with native granite, as in the Bank of Ireland at College Green in Dublin (originally the House of Parliament).

Iron dogs and various other cramps (they look like large staples) were used extensively to connect and hold stones together, usually set in lead. Unfortunately iron fixings used in combination with ashlar have rusted in places and caused the expansion and bursting of stone faces, despite the fact that they have been set in lead. The facade of the Royal College of Surgeons that faces St Stephen's Green in Dublin shows clear evidence of spalling in granite from rusting iron cramps.

THE FOUR COURTS

Location: The Four Courts, Dublin.
Age: 18th century, badly damaged in the Civil War of the 1920s.
Geology: Granite from the Dublin/Wicklow mountains and Portland stone from Portland Bill in Dorsetshire, England.

The Four Courts: An 18th century building, badly damaged in the 1920s.

Points of Interest: During the Georgian period in Dublin, granite cut as ashlar and used on public buildings became very common. It was quarried at many places in counties Dublin and Wicklow, including Dalkey, Kilgobbin, Golden Ball, Three Rock Mountain, Ballyknockan. It was used in an attractive combination with imported Portland stone. Granite was often used as a relatively thin facing and backed with rubble stone, sometimes brick. Irish limestones and sandstones were commonly used for the same purpose throughout the country.

Some of the 18th century granites were not quarried but selected from fields in the Dublin mountains and worked to shape quite easily because they were relatively soft. (Over thousands of years the iron in the granite oxidised.) However, the result of using already weathered granite is to be seen in the present-day deterioration of its surfaces. There has been quite a lot of replacement of older granite with new in Dublin.

Architectural design during this period conformed to an underlying classical geometry. The classical orders of Doric, Ionic and Corinthian are to be seen on columns and their capitals, pedestals, entablatures and parapets. Even very plain entrances to estates show classical proportional relationships in pier details, etc.

THE 19TH CENTURY

The Georgian period extended into the first half of the 19th century and various Classical styles continued into the first half of the 20th century with increasing skill seen in the cutting and laying of stone. The famine of the 1840s decimated rural Ireland, particularly the west, leading to 1.5 million people dying of starvation and a similar number lost to emigration. Work programmes were introduced by the Board of Works which had been founded earlier in the century to provide employment and financial relief although these schemes concentrated mostly on road building. In many places the term 'famine walls' is used to denote the high walls built around estates and elsewhere in this and the preceding period. Many of these were thin – roughly 450mm in thickness – and built in rubble stone with or without a coping.

In 1845 County Mayo had the largest area of land still under the Rundale system. From the 1850s on, the huge reduction in population meant that more land was available to landowners and fields were re-aligned and enclosed with walls. The Rundale system ceased to be viable except in a few areas.

Later in the 19th century a relative prosperity brought a boom in banking and led to the style known as Bankers' Georgian, using imported sandstones from Britain. But the style known as Gothic

Revival was also used in churches, railway stations, banks and many other buildings including estates and their entrances. Caen stone from France made a comeback in restoration work, replacing the Dundry stone of the Middle Ages and being used in newly built churches for pulpits and other carved work such as in the work executed at Christ Church Cathedral in Dublin.

As the century progressed the beautiful freehand-drawn and executed letters on stone began to make way for the more mechanical shapes which belonged to the print world – visible in every graveyard in the country. All types of stonecutting, from advanced geometries and precision laying to every variety of surface finish, was completed with precision. The quality of all cut stone became outstanding, reaching standards never again repeated. It is to be seen everywhere – in churches, banks, insurance offices, public and private buildings and in simple boundary walling. This was also a period when many public buildings were restored – the cathedrals of Christ Church and St Patrick's in Dublin among them – sometimes with regrettable results.

SAINTS AUGUSTINE AND JOHN CHURCH

Location: John's Lane Church, Thomas Street, Dublin.

Age: Begun in 1862.

Geology: Imported red sandstone, Irish limestone, granite and a red sandstone repair mortar used in the 1980s.

Points of Interest: Designed in the Gothic style by Edward Welby Pugin and George Ashlin in 1862, the church was not completed until 1911. The statues of the apostles were carved by the father of Padraig Pearse. During this period both the classical and Gothic styles were in vogue, each with its own strong advocates. Variations occurred which would look at home in Venice, Paris, Florence, Rome and Egypt. Great inventiveness was shown in details such as carving and it is obvious that craftsmen were allowed a certain degree of freedom in the execution of their work. The O'Shea brothers, stone carvers from Cork, excelled in carving Portland stone and their work can be seen at the Museum Building in Trinity College, Dublin, and elsewhere, including the Oxford Museum in England.

Saints Augustine and John Church, Dublin: Built in 1862 in the Gothic style.

THE 20TH CENTURY

The 20th century initially showed surviving evidence of the skills developed in the 19th century – the rebuilding of O'Connell Street in the 1920s after the Civil War, for example. Even as late as the 1940s the extension to the Munster and Leinster Bank in Dame street, now the Allied Irish Bank, shows work equal to the original building of the 1870s.

However, the second fall of the Roman Empire occurred in the 1950s when what were in essence Roman technologies – not only in stone but in many other areas of the building industry – disappeared rapidly. Over the following two generations a break occurred in the link of past and present. New methodologies appeared, many of which have caused problems when used to conserve or repair old buildings as they are appropriate to the past two generations only. There is now a need to redevelop what was lost so that we can maintain what we have inherited.

Today the dimension stone industry is reviving throughout Europe and particularly in Ireland in corporate and public buildings because of the development of more cost-effective technology that allows natural stone to be used, if not structurally, then for kerbs, paving or cladding or decoratively. Stonecutting by hand is usually now reserved for decorative carving and one-offs and the taking of a stone 'out of twist', for example, which took stonecutters so much time in the past, is now accomplished in minutes by machine. Surface finishes can be achieved using hand-held, compressed air hammers with various chisel and bush heads, or even automatically, while round-the-clock, electronically-driven, diamond-tipped saws can cut huge blocks of stone – though these methods are more appropriate for modern style buildings.

Stone walling, although quite different, is also making a comeback and it is hoped that this book will encourage that process to continue and expand, based on sound traditional principles. There is still a place in wall building for those who work by hand. It is economical to carry out basic cutting by hand for ordinary quoin stones and for general rubble and, as far as I know, there is no machine available to build walls.

CHAPTER 2

GEOLOGY

The stone used in wall building generally reflects the underlying geology of that particular area. Indigenous stone is best used where it is found because it is usually the cheapest stone to use and because, for some reason, where stone is imported into a district to build walls, such as sandstone in a limestone area, it never looks quite right.

Ireland is diverse in its geology – as any detailed geological map shows – and the traditional ways of quarrying, cutting and laying these stones are equally diverse.

Stone is classified under three main headings: igneous, sedimentary and metamorphic.

IGNEOUS STONE

Granite

Igneous stone originates from a molten magma mass and the main igneous stone in Ireland is granite. It cooled slowly beneath the earth's surface and in doing so formed a structure of large intergrown crystals which makes it ideal for cutting and building.

In Ireland it is found mainly in the Dublin and Wicklow mountains where it is a silvery colour – though at Ballyknockan in Wicklow its distinctive colour has earned it the nickname 'brown bread' – and in counties Wexford, Down, Galway and also in Donegal where it is found in delicate shades of pink. Quarries in the Dublin and Wicklow mountains, Carlow, Wexford and the Mourne mountains area in County Down have a long history of both quarrying and cutting granite. By observing how the mica – a soft mineral contained in the stone – lay, workers in the quarries could detect a grain in the granite which they used to their advantage in work involving plugs and feathers. In County Down they call it the

☐ limestone	⌇ basalt
■ granite	▥ gneiss
▤ shale, sandstone	▨ sandstone (n.r.s.)
▦ sandstone (o.r.s.) conglomerate	▧ sandstone, slate quartzite
◈ sandstone, slate greywacke	C coal
◊ schist, quartzite, marble	

'rede' and it lies north to south in the Mourne mountain range.

Granite dimension stone is a high quality stone. Unlike many rubble stones it is capable of being cut and shaped to accurate dimensions. It is used for cladding, the monument trade and all forms of decorative and facing work. Today granite dimension stone quarrying is mostly carried out in County Wicklow.

Granite consists of the following minerals:

Quartz
A hard mineral, harder than steel and the reason why granite is tough on tools and machinery.

Feldspar
Another hard mineral that gives granite its colour – white, pink, red, grey, green etc.

Mica
Both black (biotite) and white (muscovite) in colour, soft and layered like leaves.

Basalt

Basalt is another igneous stone, most commonly found in County Antrim and most famously in the Giant's Causeway. Like granite, basalt originates from a molten mass, but because it reached the earth's surface through volcanic activity and cooled quickly it developed a finely crystalline structure and short dimensional lengths, sometimes polygonal in shape, as at the Giant's Causeway, which makes it more difficult to work with.

Basalt is not used as a dimension stone but can be seen in boundary walls and farm buildings throughout Antrim.

SEDIMENTARY STONE

Limestone

Limestone was laid down layer by layer as a sediment rich in calcite, its main mineral, in warm tropical seas about 330 million years ago. The calcite originated largely from the bone and shell structures of sea organisms. As can be seen from the map, the whole central plain of Ireland and parts of the west are underlaid with carboniferous limestone. Perhaps the most famous area of limestone pavement in the world is in the Burren in County Clare.

Sadly, slabs of this rare geological phenomenon have recently been dug up and exported to England and the Continent where its light grey colour and fissured appearance is popular in rockeries. Locally it has always been used for building boundary walls.

The beautiful, relatively pure, white limestone around Cork city was extensively quarried in the 19th and early 20th century to produce dimension stone and the city has many fine examples of its use for mouldings, decorative details and carvings. The Irish dimension limestone available today is hard and grey in colour with a hint of blue. When polished it turns jet black and has been wrongly referred to as marble – Kilkenny's famous 'black marble', for example, is actually polished limestone. Most limestones when broken display a black surface area. In a few years this normally turns to grey from the atmosphere.

Limestone is used in cladding, the monument trade and all kinds of decorative and facing work. Dimension limestone is presently quarried in counties Kilkenny, Laois, Carlow and Roscommon.

Rubble walling utilises all varieties and qualities of limestone. A limestone called calp, which is inferior to the dimension stone varieties, is found around Dublin. It is a hard stone and difficult to cut or shape with easily detectable bedding planes. It was used extensively for rubble walling, and also behind the ashlar stone facades of many of the better Dublin buildings. Calp also has a high mud content and the minerals silica and alumina, and so traditionally was burnt to produce hydraulic limes which have an hydraulic set or the ability to set in wet conditions or in water. These were useful for building below ground level or in constructing piers. However, calp produces limes of variable quality and not all are hydraulic.

Sandstone

Sandstone is another sedimentary stone formed on land or in sea water from quartzite and other minerals which have been weathered out of earlier rocks. Its grains may be held together by mud, carbonate, silica or the iron which gives it its characteristic colour. Conglomeratic sandstones contain large visible pieces of other foreign rocks and look like a natural concrete. Most red sandstones formed on land in hot desert conditions.

Sandstone has not been quarried for many years in Ireland though buildings and walls in Wexford, Waterford, Cork and Kerry often used it. Old red sandstone, quarried locally, can be seen in buildings near the south coast of Ireland while new red sandstone was used in the north of Ireland around Belfast. Many red sandstones were imported in the 19th century from England and Scotland.

Shale

Shale is laid down in sea water, is composed mainly of clay and contains minerals such as fine quartz and mica. Black carbonaceous shale is formed from mud and sea organisms mixed and compressed together and is found near coalfields. It is not a commonly used stone and is never found as a dimension stone, since it is soft and is laid in very definite sheets or layers which break up quickly when exposed to the elements.

METAMORPHIC STONE

Metamorphic stone is sedimentary or igneous stone that has been subjected to heat and pressure, bringing about a metamorphosis or change in its structure (see below). Where bedding is detectable, the stone should be laid with these beds horizontal in the wall. It is frequently used as a dimension stone, in decorative work, in interiors and in the monument trade.

Originally	Now
Limestone	Marble
Shale	Schist
Mudstone	Slate
Sandstone	Quartzite
Granite	Gneiss

RUBBLE WALLING STYLES

Rubble walling goes back to the very first use of stone since it involves locally available stone that is either uncut or roughly shaped with a hammer. It accounts for the bulk of stone used in construction, being used not just for boundary walls and buildings, but as the material enclosed behind the facades of many buildings which have been rendered with a sand and lime mix. Rubble walling is, therefore, a very broad classification and is applied to a range of work considered inferior to ashlar work. The following are some of the more common styles to be seen:

Random rubble built to courses.

Random rubble built to courses

Increasingly popular from the 19th century onwards and very popular today, but unfortunately now often face-bedded and filled with concrete. The height of a course – a layer of stones laid to a set height – varies, averaging 375-450mm. When built in the traditional manner random rubble coursing has the following advantages:

- Through and bond stones which tie both leaves of a wall together (see Chapter 12: Bonding) can be accommodated quite readily on top of each course height where a level surface is established.
- Courses allow reasonably accurate estimates of the number of work hours involved in each course to be established, usually on the basis of the number of lineal metres per course per day, although the higher courses will take more time.
- More workers can be accommodated on a stretch of wall – two to a course, each end of a wall. A wall that is four courses high, with the lower courses extending beyond the higher ones, can utilise a work force of sixteen, in theory at least. Perhaps it was for this reason that random rubble built to courses was used in the 1840s on famine relief schemes.
- It is an easier style for beginners to build.

It is good practice to finish work for the day at a course height leaving a flat level surface with which to start on the following day. Take care that vertical joints in one course do not align with those of the courses beneath.

In granite work, many of the random stones may be triangular or polygonal shaped.

Random rubble (uncoursed)

Practically every old mortared stone building in Ireland, including castles, monasteries and churches, is built in uncoursed random rubble. However, if you examine the work closely you will often find elements or areas of horizontal coursing. This makes it easier to break vertical joints and it is also quicker to build with stones of a similar height. Uncoursed random rubble has the following advantages:

- When well-built it is the most attractive walling style.
- It has no inherent weakness unlike random rubble built to courses which has long, horizontal bed joints and might possibly be overturned by a horizontal force.
- It can accommodate any size of stone – within reason – because it is not limited to course heights.

Random rubble (uncoursed).

Squared rubble built to courses

This style is most likely to occur at entrances to estates and on buildings. Each stone in the wall has been squared, allowing for finer jointing to be achieved. If the stone is very finely cut, it is classified as ashlar.

Squared rubble built to courses.

Snecked rubble

This type is also most likely to occur at entrances to estates and buildings and if finely cut can be classified as ashlar. Snecks or small squared stones are introduced to break joints. No more than four stones should strike any one vertical joint.

A 19th century feature often seen on snecked work in Ireland, and in particular in Dublin, is an occasionally slightly skewed perp or vertical joint off plumb with a bed. This may have had the advantage of preventing lateral movement or cracking by incorporating overhead weight into the perp and locking stones together. On the other hand it may have simply been using available irregularly shaped stones with as little waste as possible.

There are many more styles and variations on styles, particularly on buildings. In wall building random rubble built to courses is the easiest type to build, followed by random rubble (uncoursed), while squared rubble built to courses and snecked rubble require a lot of cutting.

Snecked rubble.

County Cork: limestone and red sandstone laid as random rubble built to courses, seen here as long, horizontal lines, 300-450mm apart.

County Dublin: calp limestone rubble, squared and built to courses.

County Cork: white snecked limestone rubble with punch marks.

County Dublin: finely cut, snecked limestone with skew vertical joints. Snecks are the stones that aid bonding with stones of different heights. Skew vertical joints cut away at an angle off the perpendicular are very distinctively 19th century.

CHAPTER 4

TOOLS

Traditionally, stonecutters made their own tools. They used steel with a high carbon content which could be tempered, usually in a forge with a wind bellows and using coal as fuel. Here tools were forged to shape, hardened and tempered. The tempering is achieved by first heating the tool, then cooling its cutting end by dipping it in water, rubbing it quickly on a piece of sandstone to brighten the metal and then watching the residual heat (identified by its shade of colour) in the undipped part of the tool move down to the cool cutting end. The stonecutter can monitor this process because each variation in temperature is always a particular colour – a standard colour preferred for stonecutting tools would be straw. When the tip of the tool reaches the required colour, it is dipped in water to fix the temper, and then put to stand in a very shallow stone dish of water.

After using tools for a certain period, they need to be fire-sharpened again, especially if they are being used on granite, a very hard stone. Nowadays, many factory-produced tools use tungsten, a hard element that holds its sharpness over a long period, which is inserted into the cutting edge of the tool. It is a more cost-effective and labour-saving technique, although not as comfortable to work with as hand-forged tools.

Good tools are essential to do any work and it is worthwhile buying the best available as they will last longer and give better results. Always wear eye protection when using any tools. Dust masks and even air extractors are essential when working in dusty environments.

The following are basic tools required for working stone by hand:

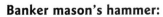

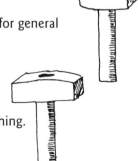

Lump hammer: 1kg in weight, for general light work.

Banker mason's hammer: about 1.25kg in weight, for pitching.

Walling hammer: about 1.75kg in weight. If the square head has unchamfered edges then it is a useful tool for quick dressing of stone.

Sledgehammer: about 4.5kg in weight.

Spalling sledge: about 4.5kg in weight.

Bull set: a type of bull set used for pitching and struck with a light sledge.

Trowel: for spreading mortar.

Line and pins: used with profiles.

Corner blocks:
for holding the line on the profiles.

Crow bar: for levering and moving heavy stone.

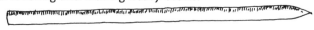

Pitcher with a tungsten insert: approx 40mm-50mm wide, heavy built, used for pitching stone to shape.

Chisel: 25mm wide for drafted margins, again with a tungsten insert.

Point: for removing waste and creating specific surface finishes.

Punch or shifting punch: with tungsten insert about 6mm wide, used for removing waste.

Steel square: for marking out such as squaring up quoin stones.

Plumb bob: for plumbing walls – when suspended it will point to the centre of the earth's gravity.

Straight edges or sighting irons: used to take stone surfaces 'out of twist'.

Plug and feathers: used to split stones to shape.

DOs AND DON'Ts

Do buy the best tools available and keep well-maintained by sharpening, fixing loose handles, etc.

Do wear safety goggles, dust mask, safety boots and gloves.

Don't allow tools such as chisels and punches to develop mushroom heads.

CHAPTER 5

SELECTING RUBBLE STONE FROM THE QUARRY

This is a key area. There are two main types of quarry: dimension stone quarries which produce top quality stone cut to shape for specific purposes such as facing corporate or public buildings; and rubble stone quarries which produce the bulk of stone for aggregates and building. Establish an intimate knowledge of your local rubble quarries (dimension stone quarries will not allow you access to the site), some rubble stone, for example, is not capable of being shaped because its bedding plane is distorted or it is not of a consistent quality. Today, working quarries with a supply of rubble stone produce aggregates for concrete manufacture, road building and so on. They use blasting techniques to extract the stone, which is later crushed and sieved to particular grades. This type of quarrying is not ideal for the production of stone for building, but it is generally the only available option (ideally stone from a rubble quarry would be quarried by hand, but unfortunately this is not a viable economic option).

All quarry owners are reluctant to allow access to their quarries for good reason as quarries are potentially dangerous places and safety is a major consideration. However, a quarry owner may allow access to those who he considers to have sufficient experience or knowledge of stone. If he does allow you access you can select specific sizes and shapes to suit your purpose. Sometimes very good rubble stone can be selected after a blast – usually stone that was not in the direct vicinity of the explosion. Stone may develop micro-fractures as a result of blasting and so be unsuitable for use in buildings. The process of extracting stone from such quarries is not quarrying, but selecting available loose stone from a quarry.

The only alternative to selecting stone yourself is to have rubble stone delivered directly from the quarry, but since it will have been selected by a loading shovel it will contain a lot of waste.

Limestone calp quarry, Dublin: the natural bedding planes are clearly visible.

Once you are familiar with your local quarry, you may find that different areas are best for particular types of stone. The natural jointing between beds in rubble quarries is often quite small, perhaps no more than 200mm or 300mm, and dictates the maximum height of stone displayed in the wall when naturally bedded. Dimension stone quarries, on the other hand, may have beds up to 3m high of top quality sedimentary stone.

For general rubble walling five different types of stone are required:

Foundation stones

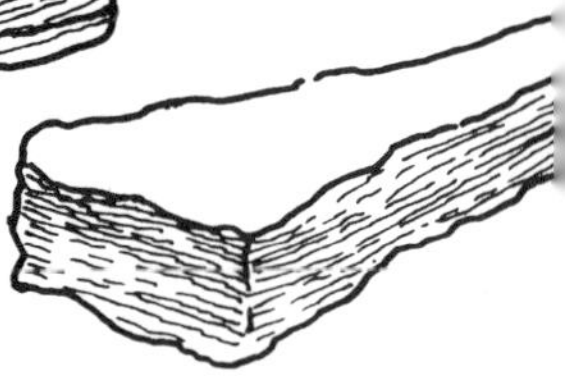

Long flat stones – any type of stone may be used – are required. If they cover the full width of the foundation, all the better, but they may be excessively heavy as a result. Generally a stonemason will use the largest stones for foundations as this means these heavy stones do not have to be lifted into place higher up on a wall. Foundation stones should lie stable without rocking or any kind of movement.

Quoin stones

Square-shaped stones are required, reasonably large in size, with faces roughly at right angles to each other and their beds, to avoid excess cutting later on.

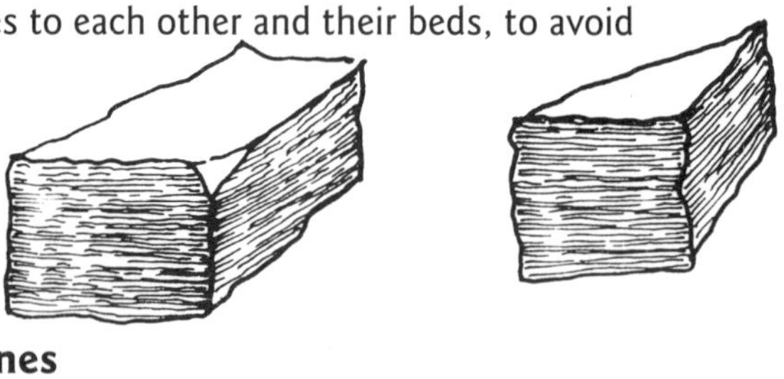

Through and bond stones

Through stones will have a length equal to the width of the wall. Bond stones will be about two-thirds the width. Laths can be cut to the length of the through stone needed, to assist selection in the quarry but normally laths are not necessary.

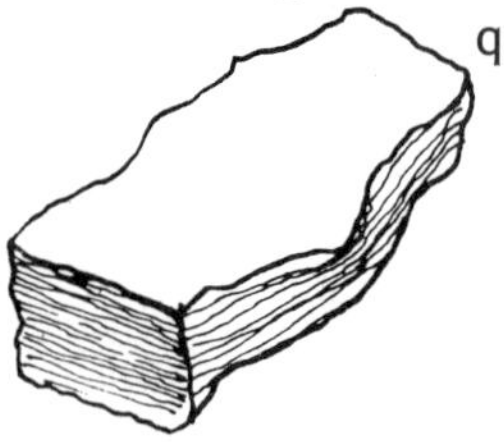

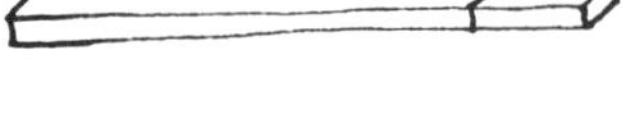

General rubble face stones

When selecting from sedimentary stone look for a top and bottom bed that run parallel to each other and also for a reasonable face. A variety of sizes will be required and it is critical to remember to select small and also flat, thin stones.

Coping stones

These will have a specific width, height and thickness. A marked lath may be useful in the selection process.

Tools for quarrying:

- **Crow bars:** at least two, each about 1.5m long.
- **Sack trolleys:** with solid wheels (if a reasonable ground surface exists).

DOs AND DON'Ts

Do work in teams and help each other in lifting and carrying.

Do select stones with top and bottom parallel to each other if possible.

Don't concentrate on selecting only the biggest stones in a quarry. Relatively small stones are quite adequate for the bulk of a wall.

Don't pick stone from lots of different locations in the quarry. This just makes collection and transport more difficult.

Safety

Because stone is such a heavy material, certain precautions must be taken in handling it. Firstly, steel toe-capped boots are a must – even a relatively small stone slipping out of your hands or off a wall will do serious damage if it lands on your toe.

When breaking, cutting or shaping stones safety goggles should be worn as slivers of stone can be as sharp as razor blades. They should also be worn when raking out or slaking lump lime where there is a danger of being splashed in the eyes.

Gloves prevent fingers getting pinched and cut by stones. A safety helmet should be worn especially when quarrying but also when there is any danger of being struck on the head such as working under scaffolding or when anyone else is working at a higher level than you.

Dust masks should be worn as required when dusty conditions exist. The dust when cutting sandstone and granite is particularly harmful if inhaled and can cause silicosis of the lungs. Finally, you should learn how to lift correctly so as not to put undue strain on your back. Seek assistance from others when lifting a stone beyond your capability. Walk, slide and twist stones which are large rather than lift them.

CHAPTER 6

TRANSPORTING STONE FROM THE QUARRY

Usually, your order of stone is delivered directly to the site. But if you have access to a quarry, then you may need to organise transporting the stone back to the site yourself. Quarrying your own stone and arranging your own transport can work out to be several hundred pounds cheaper than having a load delivered. The following is a reasonably cost-effective method of selecting and transporting stone that works in practice.

At the quarry, pile the stones in heaps, keeping the different types separate if possible. This is not always practical and sorting out may have to be done back on-site. However, copings, which are usually quickly identifiable, can easily be separated out at the quarry.

It is not advisable to leave the stone you have selected lying about in the quarry for any period of time: it may be in the way and have to be shifted, or it may be removed by somebody else looking for good quality stone. The ideal scenario is to hire a driver and a heavy truck with a steel-framed body and have the quarry loading shovel heap the stone in to it. It can then be taken back to your site. The problem is that trucks which can take the punishment of having stone dropped into them from a loading shovel are generally hard to find. Obviously, the ideal trucks are those working within the quarry but they are not usually insured or taxed to operate on public roads. Your best bet is to ask at the quarry or locally about a truck. You will have to pay for the truck, the loading shovel and the quarry stone.

For transporting small quantities of rubble stone a waste disposal skip is extremely handy. If you are lucky the drop-off truck may wait while you load the stone. The skips come in various sizes, are steel-framed and are usually low to the ground which makes them easy to load by hand. Keep in mind that you will need easy access to the quarry and a level surface within the quarry. In calculating how big a skip you need, factor in the following: Irish limestone and granite both weigh about 2.7 tonne per cubic metre. A 5m³ steel-bodied waste-disposal skip should, in theory, hold nearly 14 tonne, however, in practice a deduction in overall weight must be made for the space lost when the load is made up of loose stones. How much of a deduction you make will vary according to the size and shape of the stone being loaded.

DOs AND DON'Ts

Do load the skip uniformly to avoid tipping, or stone falling off the back of the truck during transport.

Do off-load stone as close as possible to the work area to avoid any unnecessary handling.

Don't leave selected rubble stone in the quarry any longer than is necessary.

A loading shovel in a limestone rubble quarry in County Limerick.

CHAPTER 7

SITE ORGANISATION

Assuming any necessary cutting of stone is complete, your profiles have been erected and the foundations laid, you can now organise the remaining stone to build the wall. As stone is heavy, it is necessary to place it where it will not have to be moved again and again. The diagram on the following page shows the best organisation for building a mortared wall.

GENERAL POINTS

- Keep all material back at least 600mm from the face of the wall to allow you freedom of movement.
- Place aggregates, lime, mixers and so on close to each other and convenient to the wall-building operation. They may have to be moved as the wall progresses.
- Check all materials regularly and top them up at the same rate as they are being used on the wall face.
- The organisation of stone for dry stone walling is somewhat different. Stone is placed both sides of the wall as before but with the largest stone nearest the wall and the smallest furthest away. This allows all the large stone to be used at the bottom of the wall, medium-sized stone at the centre, and the smallest stone under the coping. Through stones, quoin stones and coping stones are all set aside from the main volume of facing stones and hearting.

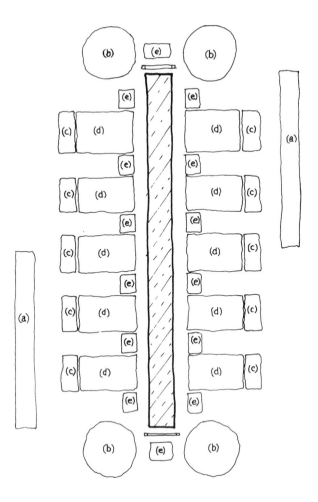

(a) Coping stones: These are selected first according to the shape and size required and placed well out of the way because they will be needed last. Stack them on one or both sides of the wall. Selecting the coping stones will be easier if they have been sorted out at the quarry.

(b) Quoin stones: Quoin stones are placed at either end of the wall and at any openings as required.

(c) Through stones: Through stones are placed behind the general rubble but clearly separated from them so that they will not be used for other purposes. Through stones have a length equal to the width of the wall.

(d) Rubble face stone: Rubble is placed as shown. This constitutes the largest category of stone required.

(e) Mortar boards: Mortar boards are placed as convenient along the length of the wall on both faces.

CHAPTER 8

BEDDING STONE

The following applies to sedimentary stones such as limestone, sandstone and shale and to a range of metamorphic stone such as mica schist, quartzite and slate.

Natural bedding

Sedimentary stones (see Chapter 2: Geology) often display bedding planes which originated from their formation millions of years ago as sediment in sea water, or as particles of weathered rock laid down on land in desert conditions.

When originally formed these beds were horizontal. The term 'natural bedding' refers to stones that have identifiable bedding which is laid in a wall in a horizontal position. Sedimentary quarries may display contortion from earth movements which took place after their formation, even forcing once horizontal beds into a vertical position. The visible beds as laid down originally should still be laid horizontally.

Dimension limestone may display no observable bedding. In the quarry, beds are identifiable, but once removed from the quarry bed and processed through machinery, they can be very difficult to see.

Edge bedding

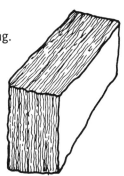

In stone walling this occurs only on the coping. On buildings, sedimentary stones such as sandstone are edge bedded on window sills, steps, copings, parapets and so on, because of the danger of the stone weathering and de-laminating layer by layer.

Face bedding

Face bedding should never occur in stone walling but, unfortunately, modern-built stone walls often face bed every single stone in the wall. The result looks like vertical crazy paving. Structurally, stones are meant to take their weight on their natural beds. Some walls laid this way display de-lamination of the beds from weathering (see Chapter 24: Bad Habits in Modern Stone Walling.

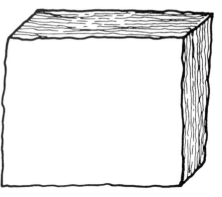

DOs AND DON'Ts

Do lay stone on its natural bed including hearting or corefill.

Don't face-bed stone.

TYPES OF PROFILE

Profiles are an essential part of wall building – they determine the line, height and range of a wall. Any time spent in the construction of profiles will be more than repaid in increased efficiency and accuracy gained by using them. However, constructing profiles can demand imagination when you are confronted with unusual circumstances such as limited availability of material and so on.

For rubble work built to courses profiles are indispensable, allowing course heights to be marked off each side of the profile. The plane, flat-surfaced, in-range finish on some old work testifies to profiles being used in the past.

Simple profile

This simple profile can be moved along a wall as construction progresses. The brace at the rear can be hinged for transport purposes. A plumb bob is dropped from the centre top and allowed to hang so that it strikes centre bottom (see Chapter 10: Plumb Bobs). Any disturbance of the profile is easily noted. The sides of the profile, shown here, are battered. A rock – as shown here – or a peg driven into the ground holds the profile in position. Use the top centre nails to sight off into the distance and establish these points on fixed objects. This is useful if profiles get knocked over or if there is a dispute about boundaries.

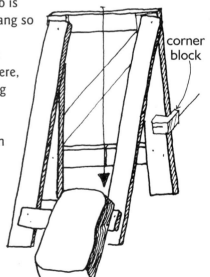

corner
block

A corner block is shown holding the line. Check the line continually for tightness and adjust throughout the day. A dipped line, particularly on coursed work or on wall tops, looks bad.

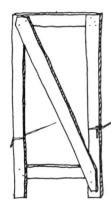

Door frame-type profile

A profile, similar to a door frame, used on a wall with two plumb faces.

Intermediate profile

A profile used at some point along a wall other than an endpoint. The centre nail on top of the profile is critical. This is sighted from end profiles and gives height and line. A plumb bob is dropped from the centre nail to mark the centre of the wall at the bottom of the profile. The width of the wall is marked off this centre.

Straight-edged profile

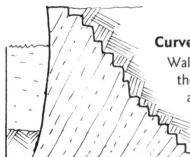

Profiles in the form of straight edges, nailed to a wall with packers behind to allow insertion of the pin which holds the line. Cut indents (niches that allow a new wall to be inserted into an old wall) in the wall first.

Curved profile

Wall faces, within reason, can be built to the shape of a profile. Curved profiles allow concave walls to be struck – as in a dam or the banks of a canal.

Battered profile

A battered profile used on a retaining wall.
The first profile is measured accurately for
batter (the inward incline of the external face of a
wall). Subsequent profiles are 'boned in' or placed
parallel to this by eye.

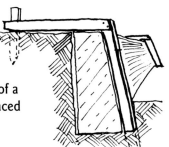

DOs AND DON'Ts

Do use profiles on all possible occasions.

Do check profiles regularly for alignment and plumb.

Do keep the string line above the work in progress
always.

Don't allow string lines to slacken.

Don't allow stones to touch or push string out of
alignment.

CHAPTER 10

PLUMB BOBS

A plumb bob is an ancient – and effective – method of establishing both plumb and level in building work. If it is suspended from a string line a plumb bob will point to the centre of the earth's gravity when stationary and denotes plumb. A right angle to plumb forms a tangent to the circumference of the earth and is what we term 'level'. The word 'plumb' derives from the lead (presumably from *plomb*, French for lead), used by the plumber or worker of lead which was employed for this purpose in the past.

An alternative method of establishing plumb is to use a spirit level which can plumb a single face of a wall, and so works only within a two-dimensional framework. If another wall is to be placed at right angles to this – at a corner, for example – it has to be plumbed separately. A plumb bob has the advantage of being able to operate three-dimensionally and if suspended at a vertical arris on a wall will delineate plumb for both walls at the same time.

The plumb bob appears to be the most simple of tools but it is, in fact, quite sophisticated. It takes a two-dimensional layout on the ground into a three-dimensional structure, allowing multiple operations to take place simultaneously, all following an infallible guide that can then be used as a reference point for each operation. The plumb bob therefore establishes order. A simple string line stretched between plumb points – whether they are marked as a line on a wall, a plank or a profile – allows walls to be built in range or to a flat plane horizontally, vertically and diagonally.

Simple plumb rule

A simple plumb rule with straight parallel edges and a plumb bob suspended from a piece of string from the centre top. The centre at the bottom of the plumb rule is also marked and sometimes a large staple is used at the bottom to prevent the plumb bob swinging about.

Battered plumb rule

A battered plumb rule for building battered or inward inclining walls.

Striking a point

A plumb bob dropped from an overhead point along the line where a wall might have to strike. The floor or foundation is marked. A simple plate with a hole to catch the line is inserted under the first stone laid to the marked point. The string is pulled tight and the stone is built just off this line (wire can be substituted for the string). The system is invaluable when an overhead point needs to be struck accurately. It is good practice in many kinds of walling to work out the finishing point before the first stone is laid, although this is rarely necessary in boundary walling.

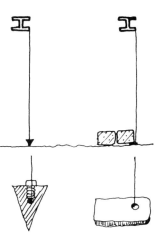

Levelling

An old system of levelling – a right angle using a braced edge is taken off a plumb rule. This was also used for sighting, its accuracy depending on the measurement of the right angle, the length of the sighting edge and the eye of the person taking the sight.

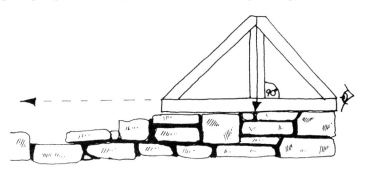

Masons' Marks

Throughout Europe and beyond a mysterious phenomenon exists on old buildings, most of them medieval, a phenomenon which can be discovered by close observation of the face surfaces of stones, and is known as masons' marks. They do not occur on all old buildings, but where they are found they are sometimes repeated elsewhere on other buildings so that the journey or work pattern of a particular stonemason can be followed.

Masons' marks are nearly always incised or cut into the face of stones. Sometimes these marks comply with an underlying geometrical pattern (as in the plan of many old buildings), at other times they are easily recognisable shapes such as a fish, etc. Ireland's masons' marks are also incised but, most peculiarly, they are also carved in relief, proud of the face of a stone.

Masons' mark carved in relief at Holy Cross Abbey, County Tipperary.

These marks were first seen in Ireland after the Anglo-Norman invasion of 1169. Their purpose is unclear, but there are many theories about them in existence. First, it gave each stonemason a mark by which he could then be identified. A son inherited his father's mark, but would make a subtle change in its layout. In a time when few could read or write, although they could design, cut, carve and build to a sophisticated level, a mark was a man's signature.

Second, there was the question of production and payment for that production which could easily be measured if stones were marked. Finally, and perhaps this is the most feasible theory, marks were used for quality control purposes. An error that occurs in the cutting process may not be discovered until the 'building-in' on-site. If a stone was marked it could be traced back to the person responsible.

It has been known today for templates (sometimes referred to as moulds, ie, top bed mould, face mould, etc.), to be altered to match a mistake which occurred in cutting, so that even if the stone was checked with the template it would appear to be all right. Only when it had been fixed or built-in on-site would the error become obvious. At this stage it

would be nice to trace the culprit. Metal files and snips (for cutting zinc templates) have been known to be banned from stonecutting workshops for exactly this reason. So perhaps this is the principal reason for masons' marks. But this raises the question of why every stone is not marked and why, at times, it is not the most important stones – those requiring a high degree of accuracy in their cutting – which are marked.

In Ireland masons' marks are sometimes elaborately carved in relief – work which would take some time to execute. It may be that waste stone was left on the face of a stone for the marks to be carved after the building was finished and the pressure of production was less. This is sometimes seen in churches built since the 19th century which have blank blocks of stone intended for future work which was never carried out.

If this is the case are these carvings in relief a mason's marks at all? Since very little is recorded on paper about stone, all that remains in the end are the buildings and we have to read these to find our answers.

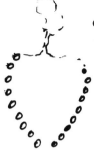

In Ireland stonemasons' marks are plentiful on edieval monasteries (but rare on castles), such as Holy Cross, County Tipperary, Kilconnell, County Galway, Kilcooly, County Tipperary and many other places.

This page: Masons' marks at Kilconnell Franciscan Friary, County Galway.

CHAPTER 11

FOUNDATIONS

Foundations are the most important part of any structure and great care must be taken in preparing and laying them. Traditionally the foundations for stone walls were laid in stone, though now concrete foundations are common. The advantage of the traditional method, which used lime mortar to build the whole structure, was that the elasticity of the mortar allowed movement to take place without any structural failure or cracking occurring, even if settlement occurred at different rates in different places in the foundations. Lime mortar is still used because of its 'give' for building in areas of the US or Australia that suffer from earthquakes.

If a wall is very wide its load might be spread sufficiently over the ground so as not to need any foundations at all – the average width of a wall varies in stone, but generally 450mm is the minimum and 600mm is the most common width. However, castle walls can be more than 4m thick.

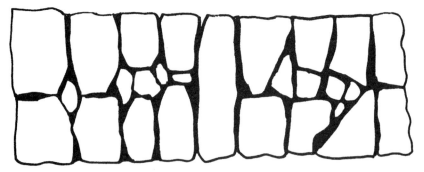

A stone foundation on plan. A key principle is to run each stone with its length running in to the centre of the wall.

CALCULATING THE EXTENT OF FOUNDATIONS

You need to decide how deep beneath ground level the foundations should be laid, as well as how wide and thick they should be. These considerations are determined by the weight of a wall and its width. An old rule of thumb for buildings was to lay the foundations to three times the width of the wall and to the same thickness as the wall. The depth below ground level was usually a set minimum, approx 450mm. Traditionally, very few stone walls were laid on foundations for buildings in this country. Today regulations on building have changed and it may be necessary to consult the building regulations and possibly a structural engineer before you begin.

Depth below ground level

How far beneath ground level you lay the foundations depends on a number of factors: the weight of the wall and any applied loads, the bearing capacity of the soil and the exposure to frost which causes the wet soil to expand – the ground is likely to experience this as 'heave' or movement. The frost-line or depth of soil affected by frost varies from country to country and climate to climate. In North America it may be as deep as 1.8m-2.4m (6-8ft) (many North American builders capitalise on having to dig such deep foundations by incorporating a basement into the building), while in Ireland the frost-line may be as little as 450mm.

Width

The width of the foundations, or how far they extend to either side of a wall, is governed by the weight of the wall relative to the bearing capacity of the soil and must also take into account the danger of the wall overturning.

Thickness

The foundations need to be thick enough to ensure that the wall does not 'punch' its way through them from above. A number of courses of stone may be used to reach the necessary thickness.

HOW TO LAY FOUNDATIONS

- Erect your profiles first – you will have one at each end of the wall and perhaps a number of intermediate profiles along the wall. If possible, do not dig and build the foundations before you have erected the profiles as variations in height or plumb will inevitably occur.
- Set the work out, mapping out the beginning and end points.
- Stretch guide lines from the profiles at each end of the wall.
- A piece of timber fixed to each profile so that it projects beyond both sides of the wall to the same width as the foundation will give a guide to the outside line of the foundation accurately.
- Make sure the foundations are laid level on good load-bearing sub-soil.
- Lay the foundations stones (see Chapter 5: Selecting Rubble Stone from the Quarry). These should be large flat stones that sit level on the soil. If a stone is slightly convex make sure the rounded side is facing upwards to prevent any spinning or rocking.
- Lay the stones in mortar which has a hydraulic additive (see Chapter 20: Lime Mortars). Stones should be laid tight against each other to prevent movement occurring. Corefilling or hearting should be packed tightly.

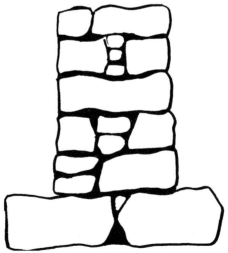

A vertical section through a wall showing its foundations.

- Keep the amount of mortar used as fill in the hearting and elsewhere down as too much will only allow excess movement to occur.

DOs AND DON'Ts

Do supervise the laying of foundations – it is critical to ensure the work is done properly.

Do use large, flat stones.

Do keep foundation stones butted up tight to each other.

Do carry out heartfilling of foundations thoroughly.

Do use a hydraulic mortar below ground level if building a mortared wall.

Don't allow foundation stones to rock – check by walking on them if necessary and pin with small stones until stable.

CHAPTER 12

BONDING

Bonding is the arrangement of stones in a wall to give structural stability and to control the placement of vertical joints. But it also affects the appearance of a wall. There are two broad categories of bonding: bed bonding and face bonding.

BED BONDING

Bed bonding, or bonding across the width and length of a wall, is essential to a wall's structural stability. A well-bonded wall reflects good workmanship, control and supervision during construction, because it is impossible to tell afterwards how well this work has been carried out. The bonding on some old walls is unbelievably good, particularly when one takes into account that the quality of bonding is not visible once a wall is completed. Tradition has it that stonemasons believed that God could see and judge their work so they took great care even in bed bonding work.

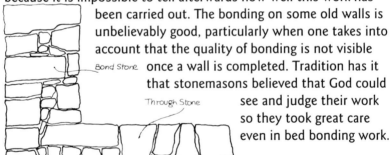

Through stones

Through stones extend from one face of the wall to the other, tying both together. They are one of the most essential elements of a traditionally built stone wall and prevent splitting of the wall, core slip, bulging and collapse. In Ireland through stones are not visible on the face of a wall, in parts of Britain, however, the tradition on boundary walls is to lay these stones so that they stick out and are highly visible.

To avoid excessive cutting of through stones before you place them, try to select suitable through stones in the quarry. Use one through stone per square metre (traditionally one per square yard), measured on the face of the wall. Through stones are laid level, though in the construction of dwellings the bed or stone itself may tilt slightly outwards to the external face of the wall so as to drain off any rainwater to the outside of the house. Stagger the placement of through stones in a wall so that they never occur directly over one another. This will give the structure greater strength.

Bond stones

Again, bond stones are not visible on the face wall. They extend a considerable distance from one face of the wall to the other, usually at least two-thirds the width of the wall. If through stones are not available, use two bond stones per one square metre of walling, measured on the face. Just as for through stones, stagger the position of bond stones and remember to place them on alternate sides and not just on one face of the wall.

Quoin stones

Quoin stones are square stones which are used at corners, wall ends or door openings and have had special care taken in their cutting. They should have their vertical joints well broken and not in alignment over each other.

Internal corner stones

Internal corner stones should extend into the width of the wall as far as practical.

Face stones

Face stones make up 95% of the stones you see on the face of a wall, and comprise virtually all those stones that do not come under any of the above headings. As far as possible, face stones should have their length running in to the wall to improve structural stability.

Corefill

Corefill is an essential but often neglected part of the structural stability of a wall. Do not use rubbish material, concrete, or sand and cement as corefill or heartfill for this purpose and if you are using sedimentary stone lay it on its natural bed. A stone should not be dropped into a wall on its edge, as the pressure of the works above will force it to act as a wedge, leading to lateral pressure and, consequently, bulging in the wall.

If possible, pack every space with stone using the minimum amount of mortar.

DOs AND DON'Ts

Do select through stones and set aside before starting to build a wall.

Do use at least one through stone per square metre of wall, measured on the face of the wall.

Do run the length of each face stone into the wall for strength.

Do keep corefilling the centre of the wall in pace with the face sides.

Don't build one side of the wall in advance of the other, or it will not be possible to bond it properly.

FACE BONDING

Face bonding is mainly about controlling vertical joints in a wall to increase its structural stability. It is what you see – the pattern and style of a wall – and the quality of workmanship involved is therefore easily assessed. There are many different types of face bonding. The most popular styles in Ireland are random rubble coursed and uncoursed (see Chapter 3: Rubble Walling Styles) which, though called 'random', involve a highly skilled process that must also take into account the breaking of vertical joints and structural integrity. Snecked and squared rubble laid to courses are common in buildings – particularly churches, entrances to country estates and railway work – dating from the 19th century, a period that showed great control of materials and a superb quality of work.

The key is to break all vertical joints as often as possible, and to make the overlaps above these joints as long as possible. As a guideline the legendary stonemason, *an gobán saor*, said the best face bonding was 'one on two and two on one' and this should be done if possible.

For common rubble work a practical rule is 'two but not three'. In other words up to two stones but not three or more can be placed on top of each other to form a vertical joint. Common sense needs to be applied with this rule: two large stones one on top of each other would constitute too long a vertical joint and not be acceptable.

The overall appearance of a wall is affected by face bonding, and even the crudest of available stone, if laid to the above rule, can look well.

It is essential to keep joints as small as possible. Lime mortar stiffens and carbonates more effectively when mortar joints are

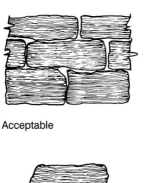

Acceptable

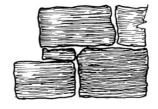

Acceptable

Acceptable

Not acceptable

reduced (see Chapter 20: Lime Mortars). To reduce the extent of mortar joints exposed to the weather small stones or pinnings are inserted at the end of a day's work (see Chapter 21: Pointing).

Another small stone seen on the face of a wall is the type used to stop a larger stone moving about. These are best inserted from the back of the stone where they will not fall out – although this is not always possible. If laid correctly they come under compression from the stones laid over them and so lock into position.

In mortared walling the pattern and size of stone visible on the face of the wall is kept uniform throughout. On dry stone walls the usual practice is to use smaller stones as the wall ascends.

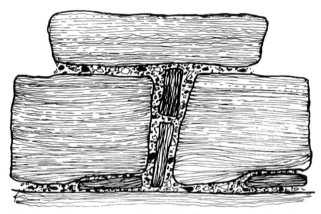

Face bonding on a mortared wall: Small stones or pinnings have been inserted between joints.

DOs AND DON'Ts

Do break vertical joints as often as possible**.**

Do keep joint sizes small.

Do pin larger stones from the back to prevent rocking if possible.

Do insert stone pinnings in larger mortar joints.

Don't exceed the rule 'two but not three'.

BALANCE

Balance is the art of placing a stone so that it looks comfortable in a wall, not just for aesthetic reasons but so that the overhead weight applied to each stone is distributed as equally as possible. All stones must be laid in balance and, if they are sedimentary, they must be laid on their natural beds (see Chapter 8: Bedding Stone).

The stones marked with an 'X' are out of balance or not laid on their natural beds.

DOs AND DON'Ts

Do lay stones horizontally and in balance.
Do lay stones on their natural beds.

Don't lay stones with a face bed length greater than their face height.

CHAPTER 13

COPINGS

Coping is the capping or covering to a wall that provides protection from the elements, animals and sometimes humans who might try to scale the wall. It also prevents small stones from becoming dislodged. Coping may have a distinctly recognisable local style. Sometimes it is the only cut stone in the wall, or the only stone not indigenous to the area, such as a barrel-shaped granite coping on a limestone wall, cut elsewhere and brought to site. Single, dry stone walls in Ireland rarely if ever display a coping of any kind.

Coping has the following practical functions:
- Protects the top of the wall from the weather.
- Prevents the lime mortar in the corefill from leaching.
- Offers vegetation little space to take hold.
- Provides privacy and security.
- Has a decorative function and gives a finish to a wall.
- Acts as a structural element in a retaining wall.
- Protects surface mortar and stone from weathering.
- Locks together the top of the wall.

If projecting it:
- Prevents individual stones from becoming dislodged.

If sharp on top it:
- Prevents or possibly frightens animals such as sheep from climbing or jumping over the wall.

Hydraulic additives such as cement may be considered for coping mortars if the wall is subject to severe weather conditions and if it is being completed during winter, in which case it needs to be kept covered until spring. Unfortunately, many modern stone walls do not have a coping because they are filled with concrete and built in sand and cement which does not wash out with the rain, although the efflorescence from salts may be a problem. However, such walls always look unfinished.

Ragged crenellated tops can be seen on some mortared walls that have lost their coping and over time these will deteriorate further. Other mortared walls appear to have been built without any coping and are now suffering the results.

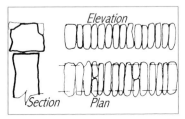

Projecting coping, level on top and with its face in range. Projections give protection and throw attractive shadows.

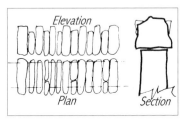

A very common projecting coping, crenellated on top and known as 'cow and calf', 'coxcomb', 'king and queen' or by a variety of other names.

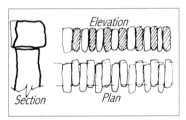

Projecting coping on alternate stones only, all of which are much the same width and level on top.

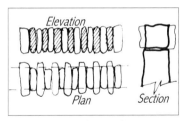

Similar to projecting coping on alternative stones (see left) but alternate stones are different widths.

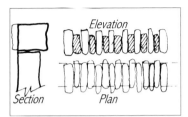

Projecting coping on alternate stones only, all of which are the same width, and crenellated on top.

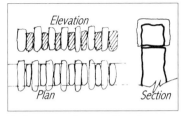

Projecting coping on alternate stones only which are different widths and crenellated on top.

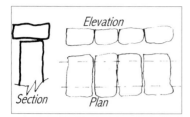

A flat projecting coping, similar to the Mourne Wall.

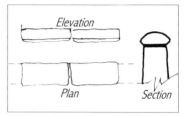

A half barrel-type coping that may or may not project, typical of those found in County Monaghan cut from sandstone.

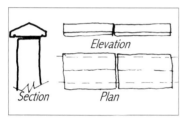

A saddle coping, which may have drips underneath.

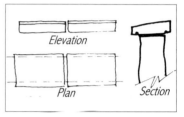

A skewed coping.

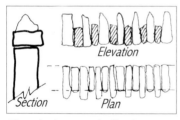

Alternate stones projecting in horizontal and vertical directions.

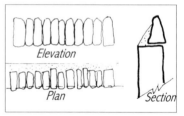

A projecting coping that may be level or crenellated on top with the back of the wall sloped off with hydraulic mortar.

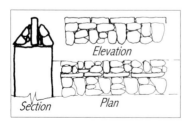

Ridgeback coping in County Meath, consisting of a stone on edge in the centre with smaller stones built on the slope at either side.

Alternate projecting coping laid in horizontal and vertical directions.

A soldier coping with indented face in South Clare.

DOs AND DON'Ts

Do build copings to match the local traditional style.

Do use a hydraulic mortar to build copings.

Do keep joints tight in copings.

Do lay a heavy coping stone on wall ends to prevent dislodgement.

CHAPTER 14

CUTTING RUBBLE STONE

For work of a reasonable quality, rubble stone needs to be cut. Traditionally, rubble stone was not cut or was simply trimmed using a hammer. Walls built with small rubble stones generally showed very little cutting and, in any case, were sometimes wet dashed or harled (see Chapter 22: Wet Dashing).

Before you cut rubble stone it is important to observe the steps carried out in Chapter 5: Selecting Rubble Stone from the Quarry. In a sedimentary quarry specific bed heights are visible, and these determine the bed heights that will be seen in the finished wall. You must select the stones you are going to cut as close to the finished size as possible. In other words, if you need a small stone do not choose a large boulder which you will then have to cut down to size.

The following instructions cover establishing a reasonable face on rubble stone and are not concerned with beds. For work of a higher standard refer to Chapter 15: Cutting Quoin Stones.

If you have selected stone with beds that are reasonably parallel to each other the result will be a reasonable cut face. However, these beds are only approximately parallel to each other so variations will occur in the width of the bed joints. The joints should fall within a certain range, and the smaller that range the better.

The cutting of rubble with a single face is done quite quickly on the ground. During bad weather you can work under cover so that a supply of cut stone is available when the weather improves.

Working the stone

- Choose the face of the stone that you are going to work, probably the one that involves the least amount of work, remembering what has been said about natural bedding. The longest or largest face may not always be the best choice – through and bond stones, for example, generally show their ends on the face of the wall.

- Position the face of the stone away from you.

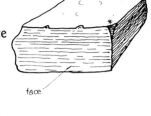

- Choose a point on the top bed in from the face edge, usually an area of damage such as a missing corner or a dip that you want to cut out of the stone (marked * right).

- Using a straight-edged bar (flat, mild steel 50mm x 6mm/8mm/10mm), scribe a line on the top bed parallel to the face arris, the same distance in from the face edge as the point you have chosen or just a little further back. Scribe the line to minimise the amount of cutting needed – a special tungsten scribing tool can be used, a sharp point or a 9H pencil. Granite workers in the past used the black carbide stick from a battery.

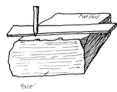

- Pitch slightly forward of the scribed line so as to leave it just visible.

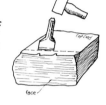

- Turn the stone upside down, so that you are working on the bottom bed, and place a straight-edged rule or piece of wood on the ground tight to the first pitched arris. Lay a second straight edge on what is now the top surface (bottom bed) and bone this in (ie, parallel the two straight edges by eye) with the bottom straight edge, moving it until it sights exactly.

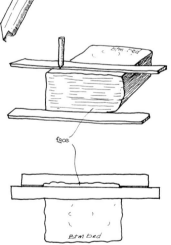

- At this stage you must consider the need to cut away any secondary damage forward of the scribed line and the need to have the face at approximate right angles to the bed so that the stone will sit flat on the wall and present a plumb face.

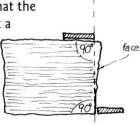

- Using stone that does not have a right-angled face to its bed can be a distinct advantage in building a battered wall as long as it is controlled. On battered walls you must batter the face of the

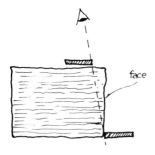

stone but keep the beds horizontal or you will end up with steps on the face of the wall. For instance, the top scribed line can be kept back to create a battered face. A wall with a batter of 100mm in 2m represents a batter of 1:20. If a stone is 200mm in height, set the second straight edge back 10mm. This can be done by eye and without measurement on rough rubble work, and the result is very attractive.

- Continue to work the bottom bed. Scribe your second line on the stone having sighted it with the first as shown above. It helps to mark the bottom line on a flat surface such as a concrete floor. Only one straight edge is then required, as the first pitched line on the stone can be put to this and a single straight-edged bar used to sight. Pitch just slightly back from the line on the stone.

- Square off all face ends so that the vertical joints on the face of

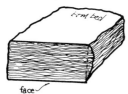

the stone will be approximately plumb. It will probably be necessary to clear any other stone along these ends that is projecting and might interfere with the next stone in the wall. This can usually be done with the hammer.

- The top and bottom pitched arrises are now connected at the ends by pitching. The face of the stone is now 'out of twist' even though its centre is projecting,

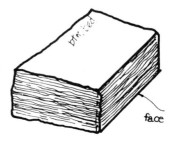

and the four surrounding arrises follow a flat plane surface. When a stone like this is laid in the wall, its bottom bed will follow the previously laid work, while its top will follow the string line.

■ This may seem like a very long drawn-out process, but in practice it is done quite quickly.

DOs AND DON'Ts

Do choose a stone that can be worked easily.
Do wear safety goggles when cutting stone.

Waste being removed by punch between drafted margins on a granite quoin stone.

CHAPTER 15

CUTTING QUOIN STONES

A quoin stone is one that occurs at the ends of walls, door or window openings, etc. It is cut from good quality stone that can be cut and shaped to a specific size. A quoin stone is cut square and in rubble work it is sometimes the only cut stone in the wall. In many Irish medieval buildings a building can be dated by the tool marks from the cutting on quoin stones at the windows and doors. Cutting quoin stones is probably the most difficult or highest skill area dealt with in this book, but it is a key area and the basis on which more advanced work is built.

Dimension stone quarries can cut and supply quoin stones using modern technology. Sometimes this work is so accurate that it has to be tooled by hand afterwards to produce a traditional-looking surface – ironic when you think how much work was once involved in cutting stone by hand to produce a flat surface.

The method below outlines one of the traditional ways of working quoin stones. This varies depending on the stone, the location and the individual stonecutter or mason but, in general, it is common to most countries in the repair and conservation of stone. It is a useful skill to have and has a definite place in the repair of stone and, in particular, in the repair of old buildings. In many circumstances working quoin stones by hand is still cost-effective and more aesthetically pleasing since it does not produce a finish with the regularity and monotony of machine-cut stone.

WORKING THE STONE

■ Select, if possible, a block of stone as near as possible to the finished size or a little larger. If a stone has, for instance, the correct bed height but is too long you can use plugs and feathers to split it.

Plugs and feathers are a very old and efficient way of splitting stone that is still used today in modern dimension stone quarries. In the Mourne quarries the technique appears to have been introduced in the 1860s from Britain and plug and feather marks are an easy way to date buildings there. Before plugs and feathers were introduced, wedges had been used and still are occasionally. Timber wedges were also used, and when wet would expand and lift a stone off its bed in the quarry.

- A series of holes is drilled in a line in the stone to be split at approximately equal distances. Two feathers are dropped into each hole and the plug is then placed between them. Each set of plug and feathers consists of a central tapered steel plug and two tapered, untempered steel feathers. They are tailormade for a particular hole diameter – anywhere from 18mm to 40mm – the length increasing with the diameter.

- The plugs are driven in one at a time in sequence across the stone to ensure they all have the same bight. It is important not to drive the plugs too hard, especially the smaller ones, or they will not last long.

- You will hear the stone splitting before you see a crack joining each hole in the line. In the past the plug holes were drilled with hand-held tools, today compressed air is used. For smaller work a heavy electric hammer-action drill with a chuck size of approx 38mm will do.

- A small set of plugs and feathers, approx 20mm in diameter and 150mm long, is adequate for producing quoin stones of up to roughly 300mm high, the holes for these would be drilled about 100mm to 150mm apart.

Stone is usually split with the grain or across or at right angles to it, though granite is easier to split if you follow the lie of the mica. (The illustration above shows plugs and feathers used on one side of the stone, though often they are used on three sides.)

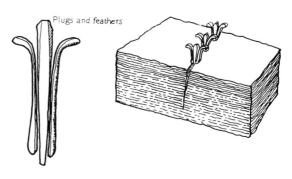

Plugs and feathers

■ Square the stone
roughly to shape.

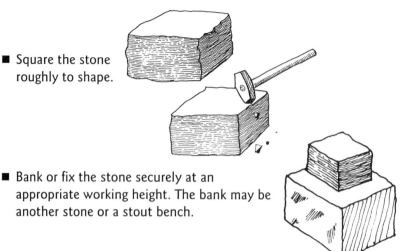

■ Bank or fix the stone securely at an
appropriate working height. The bank may be
another stone or a stout bench.

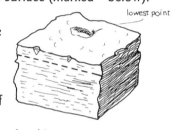

■ Position the stone so that the top bed is uppermost. Find the
lowest point on its horizontal bed surface (marked * below).
Generally this can be done by eye
but if you are uncertain use a tape
and a straight edge, or a dropped
square from a straight edge (a
dropped square is similar to the
ordinary square except that one of
the arms can move up and down
and is therefore good for measuring depth).

lowest point

■ Mark the lowest point
on the face of the stone
at one end and scribe
across the length of the
face, keeping below any
secondary depressions
or damage on the bed.

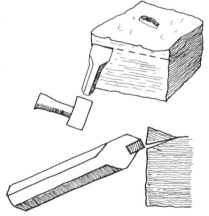

■ Pitch just above the
scribed line.

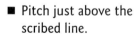

- Cut a draft or a margin to this line using a 25mm chisel. Remove with a punch any excess stone that might damage the chisel. It requires skill to cut the draft accurately. Place a steel straight edge on the draft 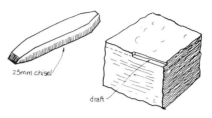 to check for accuracy. It should sit without rocking and without showing any light underneath and this takes practice to accomplish. A rusted steel straight edge is useful because it will leave a mark on any high points of the draft that need to be reduced.

- Sit a straight edge securely on this draft.
- Stand on the far side of the stone and sight the near unworked edge with the straight edge.
- Assess the lowest point on this near edge and mark it at one end (marked * below).

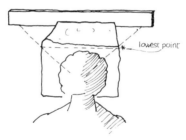

- Without moving your head, sight this mark to the opposite edge on the same face of the stone.
- Join these two marks using a scribed line.

An alternative method is to use a second straight edge and simply bone in from the first. However, sometimes you will have to hold the second straight edge against an uneven vertical face on the stone and this can be awkward.

Either method requires great accuracy or the work will have to be repeated.

For large stones this third method is helpful:

■ Cut a small table or rebate at three corners, all below any damage on the bed. Place the same size block of wood in each (say 40mm x 40mm x 40mm). Cut the last table, being careful not to cut too deep. Place the fourth block of wood in this and bone in with straight edges, gradually reducing the table under the fourth block.

■ Cut the second draft and bone in with two straight edges when finished.

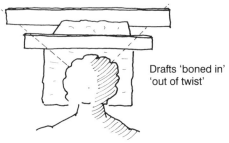

Drafts 'boned in'
'out of twist'

■ Cut the two end drafts now to connect the first two – all edges are drafted out of twist with each other and should bone in in any direction.
There are many variations on taking a surface out of twist. One is to cut the first long draft, followed by an end draft, then the next end draft is boned in and finally the last long draft is cut.

■ Remove the excess stone between the drafts by punching. The surface needs to be flat but slightly rough so that the mortar can adhere, but there is no need to continue working this to a fine finish.

■ The face is now squared off the bed, using a steel square to mark it out. The height of the quoin stone is measured down from the top bed and drafts are cut and boned in as before.

- The next bed is worked to the same method, and finally the ends are worked. Only the back of the stone can be left unworked and even this is sometimes dressed down reasonably flat.

The final finish on the stone is important and there are many variations to choose from including those shown in Chapter 16: Surface Finishes. Rubble work is often left rough or simply punched.

DOs AND DON'Ts

Do wear safety goggles when cutting.

Do bank stone securely.

Do choose a stone that can be worked economically.

Do take your time when boning in.

Don't use a chisel to remove excess stone.

Conglomerate red sandstone with granite quoins,
County Wexford.

SURFACE FINISHES

There are a huge number of surface finishes that can be applied to stone. The choice of which one is used depends on an individual stonemason's style, skills and tools, the stone available and the economics of using particular types, and the function of the finished work. Many finishes are quite distinctive and allow a building to be dated.

The following are only a few examples:

Hammer dressed: stone dressed quickly using a hammer, pitcher, bull set or similar. The stone is taken out of twist on its arrises without drafted margins. A punch is not used.

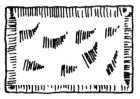

Hammer dressed with drafted margins: often used as quoin stones, which are easy to plumb because of the drafted margins.

Rough punching with drafted margins: a very popular finish.

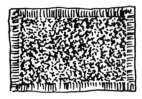

Finely punched with drafted margins: the face is finished with a sharp point.

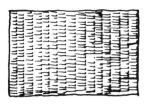

Tooled face: uses a 25mm wide (upwards) chisel – if the pattern is to be in vertical bands, then start at the top and work downwards to shed rain.

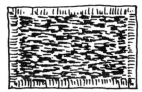

Long horizontal punching with drafted margins: a sharp point is used. Distinctively late 19th and 20th century work.

Bush hammered: in the past this was done with a point held vertically, not a bush hammer.

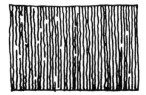

Corduroy finish/tooled face: done with a broad chisel which is worked with a wooden mallet, the furrows are aligned and wave-shaped in profile.

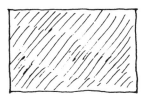

c12th century work: axe marks giving a diagonal finish. Used on relatively soft stone such as imported Dundry oolitic limestone from Bristol (1175-1400) or on sandstone. The broad axe marks indicate a method of taking a face out of twist other than using drafted margins. Notify the Department of Arts, Culture and the Gaeltacht and the National Museum if found.

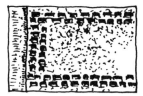

16th or 17th century work: decorative punchwork, mostly on limestones, executed with a punch about 6mm to 10mm wide, found on Irish stone. Notify the authorities if you find examples of this patternwork.

The Language of
the Irish Stonemasons

Irish stonemasons had a secret language of their own, which they used to acknowledge each other when seeking work or during the course of the working day when they wished to speak to each other privately.

The language is no longer used and went the way of so many things this century, such as working with lime mortars and building solid stone walls using traditional methods. The following is based on information I gleaned from Nioclás Breathnach over the years.

The language was called *Bearlager na Saor* or the language of the masons. *Bearla* meant foreign or strange language in Irish, though today it has come to mean English. Some words in this language are simply Irish words spelt backwards, such as *geab* for *beag* meaning small, and *lapac* for *capall,* meaning horse. Some licence is taken with words – *geab* not only means small but also bad, old, slow. A *geab lapac* is a pony or donkey.

'*Cohaec the Bearlager na Saor?*', or 'Do you speak the language of the masons?' was a question asked of someone looking for work. Unless they could speak the *Bearlager* they would not get the job or the other masons would refuse to work with them.

The stonemasons who built Cobh Cathedral in County Cork all spoke this language. The following account is taken from stonemasons building Barna Railway Bridge in County Limerick:

> *The aeis of the cín for the scíd is gone,*
> *Dioglú trithúil you'll find him,*
> *His scit and fonsúr he has girded on*
> *and the cearnógs costrua behind him.*

It roughly translates as:

> *The man of the house for the night is gone*
> *Full of drink you will find him,*
> *His trowel and chisel he has girded,*
> *and the police follow behind him.*

Even from this short piece it can be seen that *Bearlager na Saor* was not all about the serious aspects of work but also included the social and everyday aspects of the stonemason's life. The following is a stonemason emphasising the fact that he will not work for a particular employer or foreman again:

> I'd costrú the world over,
> Coidhnú in Dover, dioglú in Japan,
> With gearra coidhne I'd ceadú in China,
> Before I'd luadú for that geab man.

This translates as:

> I'd wander the world over,
> Eat in Dover, drink in Japan,
> With little food I'd die in China,
> Before I'd work for that bad man.

When work needed to be done with effort a stonemason might say to another: 'Costrú the luadar airig,' which means 'Get at the work, pal'. 'To ludú for geab poinnc,' means to 'work for small money/low wages'. 'On the costrú,' is 'looking for work'; 'costrú to Alp uí Laoire for ludú,' is to 'go [walk] to Dublin for work'.

The following is a short list of *Bearlager na Saor* vocabulary:

stonemason – *airig coda*
stone – *cloch, cáid or losáil*
chisel – *fonsúr*
trowel – *húc, lemín or scit*
apprentice – *airig flúc or geab airig*
hammer – *casúr*
church – *borabs cín* (priest's house)
cursing or swearing – *geab biniú*
big, clever or good – *trithúil*
make – *ludú*
money – *poinnc*
walk – *costrú*
work – *lúdar, lúdú*

Bloody Foreland, County Donegal: Single dry stone wall.

CHAPTER 17

DRY STONE WALLING

Building dry is a technique by which a wall is constructed using stones that sit comfortably and are balanced without the advantage of the mortar that is often used – wrongly – to achieve this purpose. These walls are sometimes only a single stone in width or constructed of sedimentary stone which is stood on end with light showing through or built of boulders. To the average stonemason involved in building walls with mortar these dry stone walls may appear to defy 'good practice', but some dry stone walls deserve to be national treasures.

Aran islands: The total amount of dry stone walls in the three islands is roughly 1,500 kilometres.

Although dry stone walls have certain elements in common with mortared walls, they are a separate entity and require a separate study. They can be quite beautiful; perhaps it is the dark shadow cast by the joints that defines each stone so clearly and displays the skill involved in its choice and arrangement, or perhaps it is the obvious relationship of the walls to the landscape.

The art of building in dry stone goes back at least 5,000 years in this country. The earliest stone walls – which also constitute the largest Stone Age monument in Europe – are at the Céide Fields in north County Mayo. Here about 250,000 tonnes of stone were used for building walls between fields in an area that covers 1,000 hectares and was used for grazing cattle (see Chapter 1: The History of Stone). There are reckoned to be about 400,000 kilometres (250,000 miles) of stone walling, both dry and mortared, in Ireland. Much of this is only about 150 years old, having resulted from the break-up of the Rundale Village system of open farming and the redistribution of the land that followed.

On the three Aran islands dry stone boundary walling between fields amounts to somewhere in the region of 1,500 kilometres. In some of the walls the smallest stones are at the bottom and the largest on top and vertical stones can be seen every so often thoughout the length of the wall. The small bottom stones are deliberately chosen to provide as few openings as possible for rabbits to get through – though of course they manage to get past.

One of the finest and most spectacular dry stone walls to be seen anywhere was built built between 1904 and 1922. It is the Mourne Wall in County Down which bounds the reservoir and surrounding land of the Silent Valley Reservoir Dam. It is 35 kilometres (22 miles) long, built of cut granite, laid dry to a height of up to 2.4m (8ft). At the base the wall is about 900mm (3ft) and battered both sides with a large projecting coping on top. This wall climbs to the top of fifteen mountains in the Mourne range and took eighteen years, working from spring to autumn, to build. The workers slept on the heather under canvas during summer to save the long climbs to and from the site each day. The fuel for tempering tools and other supplies had to be carried by hand or by donkey. Plugs and feathers were used to split the granite, and the drill holes were done with a hammer and specially tempered and shaped chisels. The evidence is to be seen on the stone themselves.

Very little information appears to exist about the wall itself, and it is thanks to W H Carson's book *The Dam Builders: The Story of the Men who Built the Silent Valley Reservoir* that I was able to gain some insight into its history.

Mourne mountains with, in the foreground, characteristic walls built of rounded granite boulder. In the hinterland 35km of dry stone walling winds around the Silent Valley Reservoir.

Each area in the country has its own style of dry stone walling which is determined by geology, skill and function. Dry stone walls have many advantages:

- Economy – they are quicker to build than mortared walls and need little or no foundations and, of course, no mortar.
- They need little maintenance and do not suffer from frost heave or attack.
- They can be built in low temperatures.
- They look more attractive.

There are a number of classes of dry stone boundary walls:

- Single stone walls – one stone in thickness.
- Double stone walls – two faces.

- Single and double combination walls with a solid base, topped with a single stone-wide wall, called a *feidín* wall.
- Retaining walls (see Chapter 18: Retaining Walls)

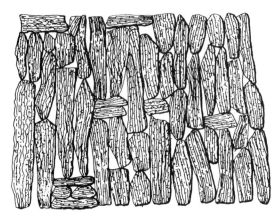

Single stone wall in the Aran islands built using carboniferous limestone.

Single stone walls

These are very common in counties Donegal and Down and in the Aran islands. They look as if they are unstable structures because of the light showing through the gaps between the stones. But this gives them an important advantage since sheep will not attempt to jump them. There is great skill involved in building single stone walls, especially if the stones are round in shape. Stones which are water rolled, round and smooth may be impossible to use since the surface tension is critical, a coarse surface will provide friction and reduce the tendency of the stones to slip and the wall to collapse.

The largest stones should be placed at the base of the wall, with smaller and smaller stones being used as the walls grows in height. A cross-section (see over) shows a pyramid shape with a gradual reduction in width from base to top.

Single stone wall in County Clare built from carboniferous sandstone.

Gates or gaps are sometimes made in these walls by using stones that can be dismantled easily and rebuilt as the need arises. In other instances long single stones are placed on end and an iron gate is hung from them. Generally these walls are limited in height because of their structure, however, in County Down the land contains such large granite boulders that they can be built into unusually high walls. Sometimes JCBs or other lifting machinery is used to lift the heaviest stones into position. On the Aran islands the walls are built of carboniferous limestone, the local stone, which is stood on its end. In most places where single stone walls occur the soil is so thin that no foundations are laid.

Single stone wall mainly found in counties Down and Donegal.

Double stone walls

These walls are built with two faces, one each side of the wall. The centre is hearted with smaller stones, which are placed carefully to fill as many voids as possible. The two faces are tied together with through stones that rest on and span each face. Through stones are a critical part of the wall. In the average wall of 1.5m high these should occur at one-third and two-thirds the height of the wall below the coping, and be no more than 1m apart horizontally. They should also be staggered over one another.

The width of the wall at its base should be half the height. Foundations should be laid in dry stone on solid ground, usually about 150mm below ground level. Foundation stones should be flat and laid tight together so that they do not rock or move.

The largest stones should be laid first. It is physically easier, and to place them higher up on a wall would cause instability. All stones except granite or similar should be laid on their natural beds and, most importantly, they should be laid with their ends showing on the face and their length penetrating into the thickness of the wall. This will give great structural strength to the wall. It is very tempting to show the largest face of the stone on the wall but this is not good

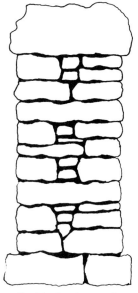

Double dry stone wall.

practice. A wall should be built to last and should not need maintenance for a long time if at all.

The coping is laid last and may vary from thin stones laid on edge to quite large, round stones, depending on the local custom (see Chapter 13: Copings).

Single and double combination walls (*feidín* walls)

Feidín wall – Aran islands.

These walls are most unusual. They have small stones at the base, which is built as a double wall, while the top half of the wall is built using large stones and is only a single stone wide. There are many theories about the origin of this type of wall. One is that it developed in southwest Scotland in the 18th century and was introduced to Galway in the 19th century by improving landlords. The solid base of the wall gives shelter to lambs while the top half, which is the traditional gapped stone, allows the strong winds prevalent in the west to pass through. At the same time the light showing through the holes makes the walls look unstable and deters sheep from jumping them. Another explanation is that the solid base prevents rabbits from getting from one field to another. One variety of *feidín* walls is plentiful around Athenry, County Galway, while another type occurs on the Aran Islands.

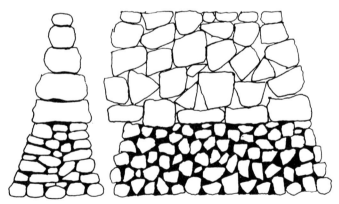

Feidín wall in the east Galway region.

In building these walls, the basic information above should be followed. A through stone placed directly on top of the smaller stones of the base and directly under the single stone wall on top will add greatly to the wall's stability.

HOW TO BUILD A DRY STONE WALL

For beginners there are advantages to starting with dry stone walling and then moving on to mortared walls. Continue local traditional practice; walls of a single stone in width may be appropriate in your district up to a limited height. Much that has been covered elsewhere in this book – through stones, bond stones, coping – is applicable to dry stone walling. The major difference is that no mortar is used in construction.

The following is a brief account of the methodology of building a double dry stone wall. The wall illustrated below has many advantages and is a very stable structure if well built:

A cross section through a dry stone wall which is 1.5m high (5ft), 750mm (2ft 6in) wide at the base and 375mm (15in) wide at the top with a coping of 300mm (1ft) high.

- The wall is set out, the profiles erected and the lines fixed.
- The top soil is removed to the width of the foundation. If the foundation stones project beyond the face of the wall it is important that they are not above ground level. This is why projecting foundations are sometimes dispensed with on dry stone walls with a wide base.
- The foundation stones used will be large, flat stones which sit without any rocking movement. These are laid tight together and all spaces between them are filled solid with stone.
- Working both sides of the wall at the same time, lay the first stones on the foundation. Use any large stones that would be heavy to lift higher up in the wall. All stones are laid on their natural beds unless they are granite or a similar type of stone

A spall inserted between two stones – always done from the back, if possible, to prevent any chance of its being dislodged.

wall, with its bed length greater than its height as discussed above. It is best to roughly grade each stone with the largest ones at the base and progressively smaller stones higher up the wall. This increases the speed of construction and does away with any unnecessary lifting of heavy stones. Each stone should sit comfortably without rocking with its face in range to the face of the wall. To achieve this it may be necessary at times to insert spalls. These should be inserted from the back, if possible, and not the front.

- Each stone – except the coping – will take the weight of overhead stones, even the small stones inserted to fill up holes. In this way no stone can be dislodged or removed from the wall.

- Through stones are critical to the structural stability of the wall and these should be placed at 1 per sq metre, measured on the face of the wall. These should be placed at approximately one- and two-thirds the height of the wall (on 1.5m high wall) and be staggered over one another.

- Corefilling or hearting is completed as the work proceeds. It is usual to fill to the top of the lowest stones before you go on to the next course of stones to be laid. It is important to carry out this process as you lay the larger stones as it cannot be completed satisfactorily later on. All hearting stones are laid flat on their natural beds and never on edge where they might act as wedges.

A coping may consist of quite thin stones except at either end where heavier stones are used to prevent collapse. Above, a wedge stone is dropped between a pair of copings where any movement occurs.

- Finally the coping stones are laid on edge. A larger coping stone is laid at the ends of the wall and at all openings to prevent dislodgement. The coping should span the full width of the wall and may project or remain flush with both faces. Any movement in individual stones can be secured by placing another small stone between it and the next coping. Sometimes coping stones that do not span the full width of the wall are used in pairs.

An elevation of a dry stone wall.

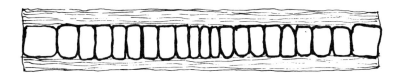

The same wall on plan.

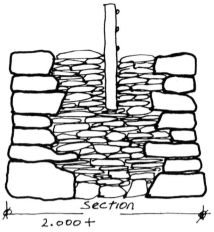

A dry stone clearance wall in Wicklow schist.

DOs AND DON'Ts

Do follow good local traditional practice.

Do use profiles.

Do lay foundations with large flat stones tight to each other and well-filled with hearting.

Do use larger stones at the base of the wall unless local practice dictates otherwise.

Do lay stones with their length into the wall.

Do heart the wall fully and as the work proceeds.

Do lay sedimentary stones and others with visible bedding on their natural beds.

Don't fill the heart of dry stone walls with concrete.

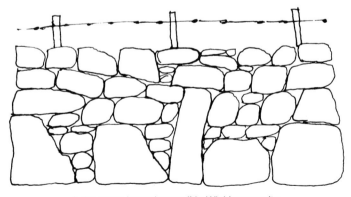

A dry stone boundary wall in Wicklow granite.

CHAPTER 18

RETAINING WALLS

Retaining walls can be seen at the sides of roads where they hold back banks of earth or between any area where a change in level occurs. The curiously named 'ha-has', on old estates are also retaining walls. High retaining walls can be seen on railway sidings, particularly near bridges, and in other civil engineering works such as near harbours. They also play a major part in landscape gardening.

Retaining walls over 1.5m high require the services of a structural engineer and are not covered here.

There are two main categories of retaining walls – dry stone retaining walls, which can be either single or double with variations, and mortared stone retaining walls.

Single dry stone retaining walls

These walls can be seen throughout Ireland, though often in a poor state of repair. They are strictly limited in height and can be dangerous if not soundly constructed. Before work starts the earth bank should be cut back to allow work to proceed safely. Different soils have different angles of repose at which they will sit comfortably without slipping any further. The volume of soil which a retaining wall holds back is the difference between the angle of the wall and the angle of repose of the particular soil. Well-drained earth banks will hold their shape permanently at an angle of 33° or 1.5 horizontal to 1 vertical. An angle of 27° which is 2 horizontal to 1 vertical may sometimes be necessary. A

Single dry stone retaining wall.

cubic metre of soil does not look like a vast amount but it weighs about 1.5 tonne. Rain decreases the friction between particles in the soil, reduces the angle of repose and is often responsible for the sudden collapse of earth banks. Safety is an issue here because if an earth bank slips it can result in the suffocation and crushing of anyone unfortunate enough to be nearby.

- Each stone in a retaining wall should penetrate in towards the bank behind, showing its ends only on the face of the wall. The largest stones should be placed at the base, although reasonably large stones are required throughout, including at the top of the wall. Many of the rules that apply to other types of walls apply here – all stones should sit comfortably and be stable, and vertical joints should be broken as often as possible.

- The face of the wall should be battered back towards the bank.

- It is necessary to backfill as the work proceeds to maintain the stability of the wall. The backfill material should be compacted and if possible should not be made up of small stones.

- Drainage is not usually as great a problem as it is in mortared walls and water should find easy access to the face of the wall. Water often seeps through the face of single stone retaining walls, especially in mountainous areas, and may form a near permanent stream flowing parallel to the base of the wall. In such cases you may need to bury a considerable portion of the base stones and lay a foundation of large stones to prevent the undermining of the wall. Better still it may be possible to construct a stone channel on top of the foundations that has occasional deeper pools with small waterfalls to slow the flow of the water and to provide places for animals to drink, plants to grow, etc.

Double dry stone retaining walls

Double dry stone walls and variations of the same are far more stable than the single stone variety. Ordinary double dry stone walls are often used to retain soil with mixed results. It is far better and safer to build walls specifically for the purpose of retaining earth.

- Before beginning work, cut the bank back so that it is safe to proceed.

- Dig foundations down to load-bearing subsoil. The foundation should be wider than the wall.

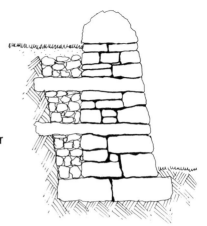

- Use large flat stones for the base of the wall that tilt slightly back towards the bank – as should all other stones in the wall. If possible the wall should be built with a batter of 1:5 or 1:6, (ie, a wall 5ft high will slope backwards by 1ft from base to top). This is greater than the 1:8 recommended for ordinary non-retaining walls.

Double dry stone retaining wall.

- Through stones are critical in this type of construction and should be as long as possible and placed as often as possible. All stones should be laid with their lengths penetrating into the thickness of the wall.

- Backfilling must take place as the work proceeds.

- Drainage is normally not an issue in dry work but special allowances can be made by leaving apertures near the base of the wall towards which ground water can be directed.

- The maximum height these walls should be constructed to is 1.5m and, like all retaining walls, they should be built by professional stone wallers.

Mortared stone retaining walls

These are capable of being built to greater heights than those built dry. Structurally, mortared walls are more cohesive (see Chapter 20: Lime Mortars) unlike dry stone walls which rely on the weight of and friction between individual stones to prevent movement. In all cases, however, the quality of work is a crucial factor and, if handled properly, dry stone can be built to reasonable heights.

There are disadvantages in using mortared stone – allowance must be made for drainage, otherwise the wall will act as a dam in wet weather or when surface water is high in the bank behind. If very rich cement mortars are used the wall may be too rigid and crack when movement occurs. For this reason vertical joints – about 10mm to 20mm in width – that penetrate from the face through to the back of the wall may be necessary every 8m or so.

The walls considered here are only 1.5m high and a structural engineer should be consulted if you wish to build higher than this. The base width to height ratio should be in the order of 1:2. A wall 1.5m high should therefore be 750mm wide at its base.

- Before beginning work, cut the bank back so that it is safe to proceed. Concrete foundations are often used for mortared walls but for walls up to 1.5m high stone is perfectly all right as long as it is laid properly.
- As before, through stones are critical to the structural stability of the wall and should be as long as possible and be used as frequently as possible.
- All stones should have their length laid into the thickness of the wall and show their ends on the face.
- A batter of about 1:5 or 300mm in a wall that is 1.5m high is about right for most situations.
- The backfilling should be of dry stone to allow drainage, and a land drainage pipe can be positioned at the back of the wall low down with access to specially constructed apertures through the thickness of the wall.
- The coping should be reasonably heavy to prevent dislodgement.
- Where ground salts are causing efflorescence on the face of a wall, the back of the wall can be rendered with a mortar that has a waterproof additive. Drainage is even more critical in this case. Dissolved salts themselves are not a problem but when they dry out on the face of the wall they crystallise, expand and cause the break-up of the face of the stone.

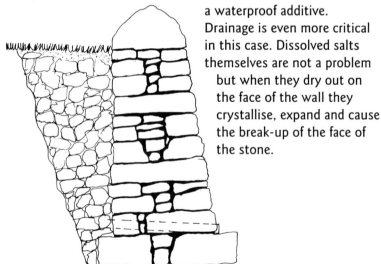

Mortared retaining wall.

A dry stone, curved retaining wall in County Kildare.

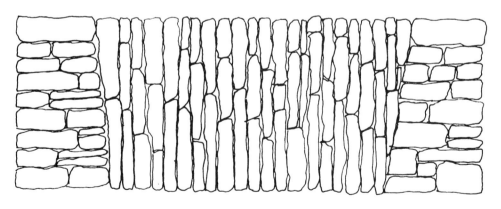

Retaining wall in sandstone in West Cork.

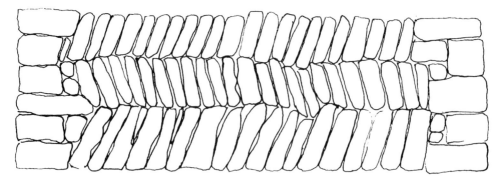

Retaining wall, herring bone style, in schist in County Wicklow.

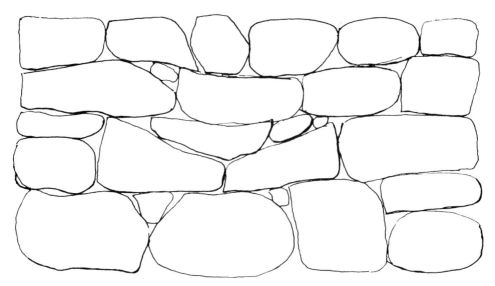

Retaining wall, roadside, County Wicklow.

DOs AND DON'Ts

Do build a solid foundation with large flat stones tight to each other.

Do run the length of each stone into the wall or bank.

Do use the largest stones at the base of the wall.

Do batter the face of retaining walls at 1:5 or 1:6.

Do backfill behind retaining walls with broken stone to assist drainage.

Do allow for drainage through walls, particularly mortared walls.

Do run through stones into the earth banks as often as possible.

Do seek professional advice if in any doubt.

Don't work where there is a danger of a bank of earth slipping and causing injury.

Don't exceed 1.5m in height without consulting a structural engineer.

Don't build the backface of a double retaining wall in advance of the front, it can never be bonded properly afterwards and if the soil slips it may knock this face and cause injury.

Ditches

Ditches or raised banks of earth, some faced with stone, are very common in Ireland particularly in less stony areas such as the midlands. Like stone walls they are an indigenous part of our landscape worth preserving. There are many different types, all serving a variety of purposes. Some consist of a wall, usually built dry, but sometimes mortared if along a

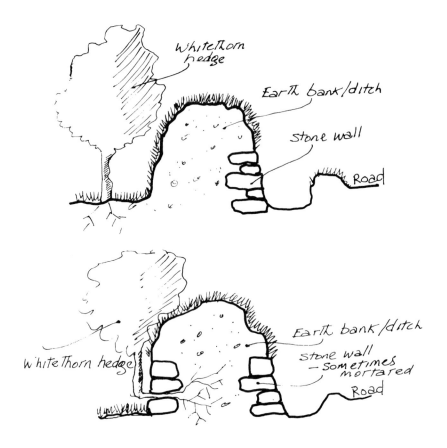

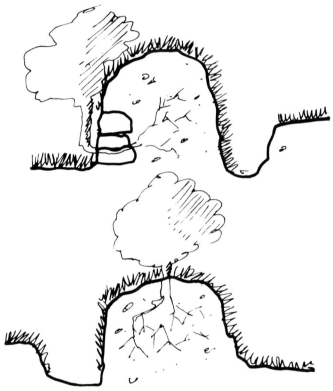

motorway. Earth is piled up both behind and over the top of the wall to give an overall height of about 1.2m (4ft). A whitethorn hedge is planted separately from the ditch or out of the face of the ditch and trained to grow vertically. The whitethorn hedge prevents stock from climbing over the ditch and also reduces wind-speed across the land.

CHAPTER 19

CIRCULAR PIERS

Pairs of circular piers, also called gate pillars and gate posts, are quite common at entrances to farm houses and fields in many parts of the country, in particular in County Antrim, County Meath – which also has examples of D-shaped piers – and the southeast. More often only one pier remains, the other having been removed to allow modern machinery access or knocked down and never rebuilt since the tradition of building piers seems to have been lost.

It is hard to date the origin of Irish circular piers. They certainly do look like Irish round towers of the 9th to 12th centuries. On the other hand in County Antrim where they are plentiful they are reminiscent of the 16th or 17th century turrets of the Scottish-style baronial castles also found there.

Circular piers whether cut or uncut often display vertical rebates for iron gate hanging and projecting stones to allow the free end of these gates to open in one direction only.

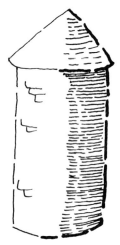

A circular stone pier with a conical top and a strong width to height ratio.

These piers are uniquely Irish and occur to a greater or lesser extent in every county but, unfortunately, they are gradually disappearing.

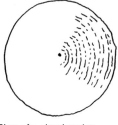

Plan of a circular pier.

To build a circular pier can be easier than building a square or rectangular one as there are fewer plumbing points (there are eight on a square pier). There is also no need to cut quoin stones as you would for a square pier, although some circular piers display all cut stone.

HOW TO BUILD A CIRCULAR PIER

The following is one method of setting out and building a circular pier. On average circular piers have a 900mm (3ft) diameter and are about 1.8m (6ft) high. Traditionally a central timber pole or iron bar set in the ground and a revolving string line were all that was used, and this is still perfectly adequate.

■ Dig a circular hole with a diameter of about 1.2m until solid ground is reached.

■ In the centre of the trench make a hole about 400mm deep using a long steel crow bar.

■ Into this hole fix a 50mm diameter steel tube which is about 2.7m long. Ensure it is plumb in all directions.

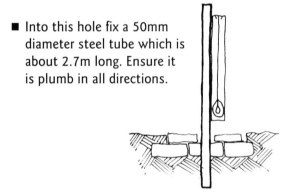

■ Lay stone foundations (see Chapter 11: Foundations) in the bottom of the trench using reasonably large flat stones. These should be laid to the outer edge of the trench.

■ Cut a plywood trammel about 600mm long and 150mm wide with a 50mm diameter hole at one end. From the centre of the hole to the outer edge of the trammel should be 450mm (1-6in), the radius of the pier.

- Place the trammel over the vertical steel tube and revolve to build pier using lime mortar. The outer end of the trammel will act as a guide to lay stones accurately to the curve. Small stones can be successfully laid on a curve without cutting. Occasional trimming with a hammer may be necessary. The centre of the pier is hearted using stone and lime mortar with care.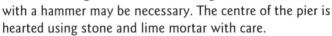

- Around 600/700mm or so from the top create a level bed from which to build the conical top. Most conical tops project about 40mm. Choose flat thin stones about 25-30mm thick. These should be cut accurately on the curve – see below on cut stone circular piers. From this base the cone is now constructed to the finished height marked on the vertical steel tube. Use small stones for this. The steel tube should be removed just before finishing.

- If appropriate to the tradition in your area wet dash and lime wash (see Chapter 22: Wet Dashing).

Circular piers with iron gates in Sligo.

HOW TO BUILD A CUT STONE PIER

Proceed as for an uncut stone pier until you have cut your trammel.

- A piece of stiff plastic or a length of plywood about 600mm long is marked out as a template to the same curve as the external face of the pier, ie, 450mm (1-6in) radius. This can be accomplished by using the revolving trammel on the vertical steel tube with a pencil held at its extremity to mark the template

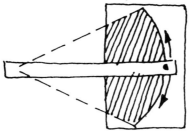

which is then cut to the marked curve. Alternatively a wooden lath with two nails 450mm apart, one fixed and the other used to score the curve on the template, can be used.

- This template can now be applied to the top and bottom bed of a stone. The stone is marked and pitched to this curve. The ends of the stone can be squared up to create tight vertical joints. The overall shape of the stone on its bed should be wedge-shaped so that its ends reach roughly to the centre of the pier. The ends of the template can be used to mark this on the stone for cutting.

- The cut stones are now laid to the revolving trammel following the rules of bonding stone on the face.

- The base of the conical cap projects about 40mm so that a template cut to a radius of 450mm plus 40mm, ie, 490mm will be necessary to mark these stones for cutting.

- Stones cut with a battered cut face are used to build the cone.

DOs AND DON'Ts

Do check the central steel tube occasionally for plumb when building circular piers.

Do lay stones with their lengths into the pier.

Do keep the central hearting of the pier filled to the level of the external face of the pier.

Do replace missing/demolished circular piers.

Don't allow circular stone piers to be demolished unnecessarily in your area. Encourage their re-building.

CHAPTER 20

LIME MORTARS

Brick and lime-making and other technologies developed by the Romans came late to Ireland. Lime appears to have been first introduced about the 7th or 8th centuries and there are fine examples of this at two early Irish Christian churches in Galway bay, Temple Benan on Inishmore and on St Mac Dara's Island. Mud or daub (*dóib*) was also used as a mortar. This was called yellow *dóib* or blue *dóib* depending on the soil type of the area. The same material was used to build mud cottages.

In the last decade or so there has been a renaissance throughout Europe in the use of lime for mortars, renders, plasters and washes on older buildings. Perhaps this is in response to the problems the indiscriminate use of rigid materials, such as cement/concrete and plastic paint on old buildings, had caused by making traditional permeable buildings impermeable and by introducing rigidity into otherwise flexible structures. Worse still many modern materials are permanent and cannot be replaced, defying the first principle of conservation: that whatever is done should be capable of being undone. This problem reappears in dealing with sand and cement pointing (see Chapter 21: Pointing).

In the past lime was usually manufactured by burning limestone, although sometimes sea shells were used. Turf or peat was used for fuel to burn limestone and bricks until well into the 20th century, the black, hard dense turf being best because it gave out greater heat. Wood was another common source of fuel and sometimes charcoal can be seen in old mortars. Coal was the most efficient fuel, of course, and used whenever available. But all fuels had to be capable of generating a temperature in the region of 900°C (judged by the colour of the limestone in the kiln during the burning).

Carboniferous limestone is common throughout the central plain of Ireland and can be found everywhere except in County Wicklow (see Chapter 2: Geology). The practice of using lime mortar brought trade to some of the areas involved – limestone was transported from the Aran islands to the carboniferous sandstone areas of County Clare, for example, and to the igneous rock areas of Connemara in return for turf which was scarce on the islands. Today lime is available readymade as bagged hydrated lime, putty lime or lump lime (see Suppliers and Services).

Advantages of lime mortars

- Allows buildings to breathe – solid walls are able to dry out more quickly when wet.
- Flexibility – structures are able to move without undue cracking.
- Repairing like with like – to use lime is to re-introduce what was originally used.
- Easy to undo – what is done can be undone, materials like stone and brick can also be recycled.
- Sacrificial – lime will fail before stone and brick whether it is used as a mortar, render or wash.
- Workability – because they are sticky lime mortars are supremely workable.

Disadvantages of lime mortars

- Education and training is required in the use of lime mortars or failures will occur.
- Problems can occur in the re-training and education of professionals and craftspeople who are steeped in cement/concrete technologies and this must be done in a sensitive manner.
- In wet/damp and cold weather the setting or, more correctly, the carbonation process of non-hydraulic lime mortars is slowed down.
- Over-enthusiasm where non-hydraulic lime is used in permanently wet conditions, etc. It will fail to carbonate properly.

In the past, many farms used lime kilns to produce lime for soil improvement rather than for building – the Irish for lime is '*aoill*' which translates as 'to manure'.

Temple Benan, Inishmore, County Galway: an early example of the use of lime mortar.

Two types of kiln were common in the past:

Mixed feed kiln

This type of kiln was still used well within living memory, the ratio of turf to stone varying probably due to the quality of the fuel and the efficiency of the kiln. As the name suggests the stone and fuel were mixed together in the kiln. This was done in alternate layers of stone and fuel. The stone was broken down to about fist size. It took a number of days and nights to complete a burning.

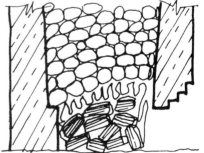

If fed from the top continually it would be called a continuous mixed feed kiln. Some firing materials may have produced a slightly hydraulic effect in the lime which, when used in mortar, allowed it to set in damp conditions.

The flare kiln

In these kilns the fuel and the stone were kept separate. A rough arch was built over the firebox in the kiln using the limestone to be burnt. This supported the rest of the stone overhead. The fire in the fire box was kept burning by adding fuel as necessary until the limestone was converted to calcium oxide (lump lime).

Environmental factors

These were important in all kilns. Heat loss – in cold weather or via the kiln walls – was doubly wasteful, prolonging the burning process and using extra fuel. For this reason kilns were built into hillsides or partly buried, which also made them easier to load. Any exposure to wind had to be controlled or it could make the kiln race and over-burn the stone. The same problem occurred in brick burning where canvas sheets in frames or, in later times, galvanised sheeting was positioned to reduce the effects of wind.

During wet weather the extracted lump lime would have to be protected. For economic reasons it was best to have a source of fuel and stone close by.

NON-HYDRAULIC LIME

Unlike cement, non-hydraulic lime cannot set under water; it sets through the process of carbonation which is the re-introduction of carbon dioxide lost in the kiln stage and it has all the advantages set out above. Although, at the moment, it is unlikely that anyone would produce non-hydraulic lime using a kiln, the process and the materials produced during each stage are of interest.

Carbon dioxide

This is a gas given off in the burning process and the limestone is converted from limestone (calcium carbonate) to calcium oxide (lump lime) in the process. The size of the material stays the same but the weight is radically altered – falling by nearly 50%.

The following explains how to produce putty lime from lump lime – a very dangerous process. In most cases, slaking and so on is unnecessary as mature putty lime is readily available (see Suppliers and Services).

Lump lime

Lump lime remains after the burning process. It is critical that it is not allowed to air slake, a process that occurs when the moisture in the air comes in contact with the lump lime, causing it to break down into a useless powder. Lump lime should be used while fresh from the kiln, either by mixing it with aggregate and water to make a hot lime mix or by converting it to putty by slaking. Lump lime can be stored for a period of time if sealed in an airtight container.

Slaking

This takes place by adding the lump lime to approximately twice its volume in water. This is a highly dangerous process, requiring special training. Protective clothing, gloves and goggles should be worn. Cold lump lime added to cold water will raise the temperature of the water beyond boiling point in less than a minute. If too much water is used the lime will drown, if too little, incomplete slaking will take place. This is a key area and the quality of the putty lime produced is dependent on it.

Slaking

Lime pit

Putty lime

Putty lime is produced from the slaking process. This can be run through a sieve into a pit and kept covered by a layer of water so that it can be stored indefinitely – in fact the longer it is stored before use the better as any small, unslaked particles that have managed to get through the sieve break down, and the whole mass becomes increasingly stiff like a soft cheese. This is lime putty at its best. Excluding air during this process, by covering the putty lime in water, for example, is critical, although too much water may prevent the putty stiffening.

Mortars, renders and plasters

These are produced by taking putty lime and mixing it with aggregates. Traditional mortars were mixed at ratios from approximately 1 putty lime to 3 aggregate down to 1:2. The amount of lime required in a mix is determined by the void space in the aggregate, the ideal being 33%.

MIXING

Traditionally mixing was done by beating the putty lime into the aggregate. A mortar mill with two revolving wheels was sometimes used for this purpose. In the Carlow/Kilkenny/Laois area single stone wheel mills were pulled by a horse and pulverised gravel into a fine aggregate while crushing and slaking the lump lime at the same time. The mortar produced was used for rendering the underside of slates. This was not the primary purpose of these wheels, which were communal – more often they were used to produce culm, a mixture of anthracite and clay used for burning in domestic fires. The wheels on most mortar mills could be adjusted to prevent the pulverisation of the aggregate.

Mortar mill

Nowadays mortar mills are beginning to make a reappearance in Europe. It is to be hoped that they will soon be common in Ireland. The mortar mill produces the best mortar by far because no additional water is required other than that contained in the lime putty which is about 50% water. Shrinkage is reduced because the amount of water in the mix is controlled and the mortar stiffens more quickly because there is less water to evaporate. This is a key area.

Where mortar mills were not available the putty lime had to be beaten in by hand, an exceptionally hard job to do correctly. This is probably the origin of the saying, 'A good labourer always uses the back of the shovel' which was often heard on building sites until recently. These days there is unlikely to be a single builder in the country using a mortar mill, although this should change with the increase in this type of work. In the meantime ready-mixed mortars produced in the traditional way can be obtained (see Suppliers).

If a cement mixer is used, additional water is invariably used in the process because the materials will stick together and not mix properly. If there is no other alternative and the mix is too wet you can use the souring out process (see below) to eliminate excess water by storing the mortar on a concrete slab under plastic and letting the excess water run off over time. This is a far from perfect solution as a certain amount of lime will be lost in the process.

Souring out

This is the process of mixing lime and aggregate well in advance of use. It produces very workable mixes and is much favoured by stonemasons, plasterers and bricklayers alike. Basically the materials are pre-mixed (see below) and stored under cover in a damp state to prevent drying out and access by carbon dioxide. The mixes can remain in this state indefinitely – six months was common not too long ago. Similarly, you can use airtight containers and store the material for future use.

Souring out allows the mixing process for a project to be done in a controlled short timespan and to be stored away. When it is used it will have a consistency in workability, mix ratios, colour and so on which the alternative – mixing when required – lacks, being inclined to produce patchy results unless controlled strictly.

Traditional mixing   'Souring Out'

Hot lime mixes

Instead of running putty through a sieve into a pit, another very common traditional way of working, especially for mortars used in rubble stone walling, was to form a depression in the aggregate, throw in the lump lime, cover with more aggregate and add water. This was turned over a number of times during the souring out process. The slaking and the mixing were therefore combined.

The result was highly workable mortars, but it contained unslaked particles of lump lime which can often be seen in old mortars. This is not a process to be used for producing finishing plasters as unslaked particles will expand and explode when introduced to moisture at any later stage. For general masonry work, however, it produced excellent mortars which are incredibly sticky. The mix was again about 1:3 but on slaking lump lime will produce twice as much lime putty so the final mix was 1:1½. Lump lime and aggregate can, of course, be mixed/slaked in a mortar mill.

Knocking up

Knocking up is the process of reworking the soured out mortar into a cohesive mass so that it is workable. It takes place just before using. Again the mortar mill is best and no water is added.

Re-entry of carbon dioxide gas (carbonation)

This takes place over time and the mortar sets thus completing the cycle. Coarse aggregate and the natural large crystalline structure of lime assist the process. This is an important stage, once we understand that non-hydraulic lime sets through the re-entry of the carbon dioxide gas that was originally lost in the burning process, that process can be assisted. It is also obvious that non-hydraulic lime mortars cannot be used by themselves in continually damp conditions such as below ground level because of the unavailability of carbon dioxide gas. These conditions would, however, be suitable for long-term storage of putty lime.

In renderings and washes, thin coats are applied to achieve fast carbonation. In pointing, mortar is applied in a number of stages for the same reason if joints are deep. In stone walls built with lime-only mixes, carbonation will occur for a distance in from the face and at a slower rate but less effectively at deeper levels. This adds a further degree of flexibility to old structures which is a desirable quality, allowing them to move without cracking. Adding cement-based concretes into these cores undoes this advantage and introduces rigidity into otherwise flexible structures leading to cracking. If the wall is part of a building then the wall's inability to dry out will also become an issue. Carbonation is a slow process in thick masonry walls so, fortunately, hardening occurs allowing work to continue.

Hardening

Hardening occurs from the evaporation or loss of water in the mortar which produces a stiffening of the mix and is sufficient for most work to continue. In winter or in wet conditions when little drying is taking place the mortar will be slower to dry. Knowing this allows us to make allowances to assist the drying-out process.

For instance, when using putty lime dry aggregates are necessary to produce stiff mortars. This is critical in mortars for stone work, pointing, etc. Some materials such as brick have a high absorbency rate and assist water loss in the mix. This is useful in winter but can be a disadvantage in summer. One of the advantages of using a hot lime mix is that very wet aggregate can be used as the lump lime will absorb twice its volume in water. In dry weather pre-wetting is normally crucial, followed by covering or protection from drying out too quickly by using plastic sheeting. In wet weather the mortar will be slow to dry and, if covered in plastic, allowance should be made for the entry of air behind the plastic sheeting to assist drying out and carbonation to occur. Cold weather can cause frost damage, and insulation will be required in addition to plastic covering.

The above is simply good practice and should not be seen as something special. Much of the above is equally applicable to cement-based mortars.

HYDRAULIC SET

A material that displays a hydraulic set has the ability to set or partly set under water. The following are all materials that give a hydraulic set:

Hydraulic lime

In the 19th century hydraulic lime was produced in some areas of Ireland by burning limestone that had a clay content and contained alumina and silica which gave a hydraulic set. It was classified into feebly, moderately, and eminently hydraulic lime. Although this is no longer manufactured in Ireland it is still produced in some European countries. The calp limestone common in the Dublin area displayed these properties but varied in quality. Hydraulic lime can be added to non-hydraulic lime and aggregate in much the same way as Portland cement.

Brick dust

Brick dust was used to create a hydraulic effect by the Romans in Britain. During the 18th century it was used in bridge building along the Liffey river in Dublin. It can be detected in small quantities in mortars and renders but is not common in Ireland. For brick dust to give a hydraulic effect it must be crushed very fine or it will simply act as an aggregate – a good aggregate nonetheless, absorbing water in the mix and allowing entry of air. Low-fired brick gives the best set. The mortar mill was responsible for crushing clay brick into powders and aggregates.

Pozzolana

Pozzolana was used by the Romans and is still used in Italy and elsewhere today. This is a natural material derived from volcanic ash which gives a hydraulic or pozzolanic effect. The Romans used pozzolana with non-hydraulic lime and aggregate to produce concretes.

Portland cement

This was developed in the first half of the 19th century and used in Ireland for external stucco work from 1850s on. It became increasingly popular in Ireland after the First World War for mortars, renders and plasters. Unfortunately the transfer from lime-only technology to cement meant that standard mixes such as 1:3 were copied, and the result was that extremely hard impermeable mixes increasingly became the norm.

Portland cement, non-hydraulic lime used as a hydrate (bagged lime) and aggregate are used today for modern building purposes and in restoration work. Portland cement gives a fast hydraulic set which means that work can proceed during most of the wet winter weather. Recent indications are that cement/lime/aggregate mixes with a cement content weaker than 1:3:12 are failing (Smeaton Report, English Heritage). A common mix for ordinary Portland cement in conservation work is 1:2:9. The problem occurs when the ratio of cement to lime in the mix is less than half.

Pre-mixed lime putty mortars

These reduce the amount of on-site work necessary in preparation and ensure a consistency from start to finish. Aggregates are specially selected and graded to suit a range of work such as pointing, from fine joints in ashlar work to coarse joints in rubble. Special mortars can be prepared for conservation work since a particular colour and shape of aggregates may be necessary to match old work. Other specialist washes and renders are available (see Suppliers and Services).

SUMMARY OF MIXING METHODS

In both traditional and modern mixing methods the souring out process produces the best mortars.

Traditional methods

Putty lime mortars

- Lime putty and aggregate is mixed in the mortar mill without any additional water, soured out for a period of time and then knocked up in the mortar mill and used.

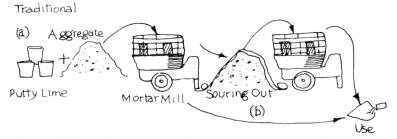

- Alternatively the lime putty and the aggregate are mixed in the mortar mill and used immediately.

- The lump lime and the aggregate are mixed together with the addition of water, soured out and knocked up when required.

Traditional
(c) Hot Lime Mix

Lump Lime + aggregate

Souring Out

Use

Modern methods

- Hydrated lime and aggregate are mixed in a cement mixer and soured out for a period of time. When required, cement is added to the cement mixer and the resulting mix is used immediately.
- Alternatively, cement, hydrated lime and aggregate are mixed together and used immediately.

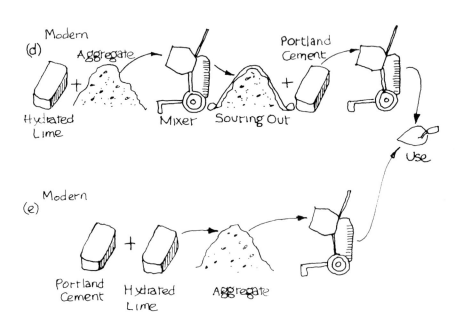

Modern
(d)
Aggregate
Portland Cement
Hydrated Lime
Mixer
Souring Out
Use

Modern
(e)
Portland Cement
Hydrated Lime
Aggregate

DOs AND DON'Ts

Do use lime-only mortars where appropriate.

Do receive training and education in the use of lime.

Do examine and **match** aggregates in mortars where appropriate.

Do give clear concise instructions to those engaged in mixing.

Do keep the water content of mixes low in most cases.

Do supervise all work in progress closely.

Do encourage the use of lime for the repair of old buildings.

Don't use non-hydraulic mortars in permanently damp conditions such as below ground level.

Don't leave recently used lime mortars to the mercy of the wind, rain, frost and sun. Cover them.

Don't add small amounts of cement to lime mortars – these mixes produce the worst failures.

CHAPTER 21

POINTING

Pointing is the replacement of surface mortars, which have been lost through weathering, in joints between stones. It prevents water entering and lodging in a structure, the wall fabric deteriorating, the loss of structural integrity and stones becoming easily dislodged. Pointing is not about making a wall waterproof or impermeable. 'Repointing' occurs when older pointing has to be replaced, while 'replacement of bedding mortars' occurs when excess weathering has occurred and a large percentage of the original mortar between stones has been lost. Pointing is the most controversial subject in the repair of stone walls and the technique discussed here refers to lime mortars (see Chapter 20: Lime Mortars for details on mixing, etc.).

Consider the following carefully before you carry out any pointing or repointing:

■ Is it necessary? Has the existing mortar deteriorated to the point where it must be replaced? You can judge this from the amount of recesses and holes, dislodged smaller stones and loose friable mortar in the wall. Previous pointing in rich sand and cement may have shrunk as it dried and pulled away from the background, allowing water to get in behind it. This is quite common where the wrong materials have been used for pointing or where insufficient raking out has taken place.

■ Mortar that is flush or nearly flush to the face of the stone and full, without shrinkage cracks, but also soft and capable of being removed quite easily, should not be replaced. Many such mortars are being replaced on the assumption that because they are soft they are somehow inferior. In fact, their softness allows movement to take place without cracking and also allows wet walls to dry out quickly. These soft mortars are often the original mortars used in the construction of the building or wall and are valuable on that basis alone. They have also proven that they can survive over time. These mortars are irreplaceable.

- Decide whether you need to point an entire wall or whether selective pointing will be sufficient. Mortars do not deteriorate consistently over the surface area of a wall. As a result selective pointing is best – replace only what needs to be replaced. For aesthetic reasons and to achieve uniformity of finish, the whole surface of a wall is sometimes pointed. This means that perfectly good mortar is removed and it adds unnecessary costs. The alternative – selective pointing with lime mortar – will look patchy initially, but after a year or two the new surface areas will tone down and blend in with the existing mortar. An over-white appearance may indicate excess water in the mix.

Selective pointing only is needed on this wall. Deterioration or loss of mortar is indicated by the dark areas shown and only these areas should be pointed.

- Corefilling may be necessary before pointing. Walls that have lost their coping and been exposed to the elements for some time may have leached out their lime mortar from the core. This can create structural instability with stones falling out and core slip leading to bulging. Pointing alone will not correct these problems and core mortars will have to be replaced. Specially designed mortars, applied with a gravity-fed system or low pressure pump, are normally sufficient.
- The degree of exposure to the elements that a wall is subjected to needs consideration, including not just the location of the wall but also the degree of susceptibility to weathering of its different components. When designing mortars for pointing, these factors need to be considered – it is not always a case of pointing a wall over its entire surface using just one mix.
- The time of year at which the pointing takes place will affect the

rate at which mortar mixes set and how much protection is required for the finished work from frost, wind and rain.

- Pointing should take place on the shaded side of a wall during sunny weather. Similarly, warm dry winds and rain can be avoided if the work is organised to accommodate changes in the environment.
- The nature of the stone to be pointed will affect the mix, whether it is weak, soft and decayed, or hard and durable. The permeability of the stone also has to be taken into account. The mortar must always be designed to be sacrificial, in other words to be weaker and more permeable than the stone being pointed. Pointing mortars which are hard crack with movement, shrink, do not allow water to evaporate, cause accelerated decay in stones from frost and salts and, worst of all, cannot be undone when these problems become evident and you want to replace them.

Once you have decided that pointing is necessary and also decided whether it is to be selective or complete, you need to carry out the following steps before you can begin pointing:
- Analyse the existing mortar in terms of grading, lime ratio and void space so that you can match it.
- Rake out the deteriorated bedding mortars from the joints.
- Brush and dampen the joints to prepare them for pointing.

MORTAR ANALYSIS

If old existing mortars need to be approximated the procedure is to take samples and have them analysed in a laboratory or conduct a simple field test, depending on the importance of the work. A simple field test may involve sieving the aggregate in the original lime mortar mix to determine grading, visually assessing the colour and shape of particles, checking the void space and determining the lime ratio in the original mix. A laboratory can do a geological analysis of the aggregate.

In the past aggregates were used without the specialised grading and batching that is available today. Larger stones were discarded and coarse variable aggregates were used for rubble work. Lime was also variable (eminently, moderately or weakly hydraulic, non-

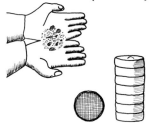

hydraulic, etc). Therefore a series of samples taken from the same wall may yield different results. A representative aggregate is developed from these results and the original lime content of the mortar can also be determined (see Suppliers and Services).

Old mortars for rubble work are generally coarse and if you are used to working with modern mortars which have fine sands you may find them a little unusual to begin with (for pointing mortars see Chapter 20: Lime Mortars).

Mortar joints in rubble stone walling

Mortar joints can include three different types of small stone:

- The larger pieces of coarse aggregates/gravels that occur naturally in the laying mortar itself.
- Pinnings, which are the relatively small stones, often flat, inserted under larger stones to prevent them rocking or moving. These are under compression.
- Pinnings or spalls which are stones inserted into large mortar joints at the end of a day's work to reduce the surface area of lime mortar exposed to the weather. These are not under

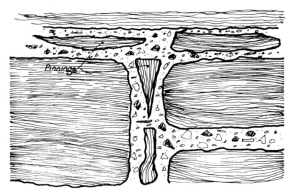

compression and are lost through weathering over time. They are often discarded during raking out and not replaced, yet they are an essential part of the visual and structural fabric of a wall and should be re-introduced during pointing.

RAKING OUT

This is the removal of existing deteriorated bedding mortars which are already missing or partly missing, friable or turned to dust, etc., and it is done using hand tools.

- Blade-cutting machinery such as angle grinders and con saws should never be used for this process. If they are used because the existing mortar is too hard to remove with hand tools then you must seriously question why the work is being raked out in the first place. Angle grinders have been wrongly used to widen the fine joints in ashlar work to allow joints to be pointed more easily. Even in the most skilful of hands blade-cutting machinery is never acceptable.

- Compressed air tools are sometimes used and at low pressure may, in some instances, be acceptable.

- Raking out should start at the top and work down.

- Use a plugging chisel for raking out. The skew blade should travel parallel to the face of the wall. This ensures control of the depth of the raking out. Often a plugging tool is used the opposite way round, with the point stabbing the mortar bed erratically and as a result, leading to an uneven depth of cut.

- Rake out each joint to a depth equal to twice its face width, except on very wide joints where individual stones could become dislodged. Joints should be raked out cleanly.

- The illustration (right) shows two joints, incorrectly raked out and, unfortunately, very common. The top one is insufficiently raked out while the bottom one displays a V-shape in section.

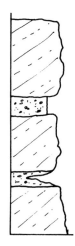

BRUSHING AND DAMPENING

After raking out, you need to brush and dampen the joints. This is done to ensure that there is no rapid loss of water from the pointing mortar through suction, leading to rapid drying out and excess shrinkage.

- Use the head of a sweeping brush or similar and brush out thoroughly since it is important that all loose mortar is removed. Start at the top and work down.

- Next dampen the wall using the spray attachment on a hose. Do not drown the wall, but give it a reasonable soaking and allow it to dry off until just damp. This is a key step which is often neglected particularly on boundary walling where the water supply may be at a distance. In dry weather dampening becomes essential as if it is overlooked the mortar will dry too fast, turn to powder and fail. In damp weather a light spray may be sufficient.

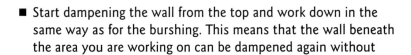

- Start dampening the wall from the top and work down in the same way as for the burshing. This means that the wall beneath the area you are working on can be dampened again without

having the fresh mortar run down the face of the wall. A portable spray bottle may be enough to keep the wall damp once it has already been soaked. During the winter, soaking the wall may create problems if there is a danger of frost. Use your common sense and make adjustments to suit changing environmental conditions.

POINTING

Application

■ Use lime mortar in a relatively dry but plastic state and compress it into place using a flat steel jointing bar. The movement (see below) is:

1. A direct push at right angles to the wall.
2. A finishing sweep to the right, ensuring that all new material is joined and compressed into previous material.

Note: A right-handed person (see above) should always work from right to left (it is not good practice to work in the opposite direction or the finishing sweep described above will not be done correctly). If two right-handed people work on the one wall face, one should start on the extreme right and work towards the left, the other should start in the centre and work towards the left. Left-handed people should, of course, do the opposite. The best solution is to have right- and left-handed people work together where possible.

- Apply the hawk (the one I use for holding mortar for pointing stone and brickwork works well on walls that have reasonably  flat surfaces). A projecting sheetmetal top (projecting by 6mm) rests on the bottom arris of the horizontal bed joint, allowing mortar to be compressed into position using a flat steel bar without staining the work underneath. The other edges of the hawk are plain so that it can be used in the normal way if required. Vertical joints are not a problem as the mortar can be lifted off the hawk with the flat bar and pressed into position.

- Watch out for bed joints filled flush – if arrises are weathered and rounded as shown filling the joint flush with the wall will result in too large a surface area of mortar showing. Therefore it is preferable to keep the joint back from the face.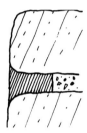

- A jointing bar that is too narrow has been used in the example shown here. Compression occurs only in the centre and any 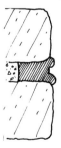 loose mortar is pushed out to either side. Therefore it is crucial that everyone working has access to a variety of bar widths.

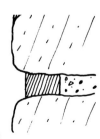

- The pointing in this case is kept behind the face so as to show the original joint width before weathering. This is considered correct practice nowadays.
- A joint partly filled and left rough on the face for filling later on. This is good practice

on deep joints because it allows partial water loss and carbonation to occur. The same principle is applied in rendering, lime washes and so on. In this case the large mortar joint is partly filled, left rough and partly or fully filled again later. Dampening down with a light spray will be necessary between applications. The timing between applications is important – depending on the weather you may apply further applications on the same day or after a number of days. The procedure is not difficult or time consuming, the work continues without stopping and the doubling back occurs as a natural process of this activity.

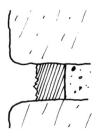

- The mortar joint has been slightly over-filled and compressed prior to beating with the bristle brush. This is good practice and prevents joints ending up being over-recessed.

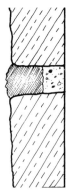

- When the lime mortar has hardened, beat it with a bristle brush to give an even surface texture and finish similar to that found on old mortars. The brush should be the same width or thinner than the joint width so that it will not stain the stone. The mortar should be reasonably hard but still flexible enough to compress and expand so that it will eliminate any minor shrinkage cracks that may have occurred (following the procedure outlined so far should keep shrinkage to a minimum). On no account should the brush be used so that it leaves long, visible brush strokes. This is very common and most unsightly.

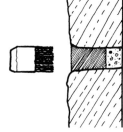

- Insert small stones (pinnings) into the pointing to reduce the exposed surface area, prevent shrinkage and assist carbonation.

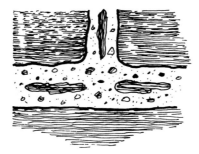

- Protect the work from rain, sun and wind. If th is a possibility of frost use insulation. Always allow some ventilation so that drying and carbonation can occur.

INAPPROPRIATE MODERN POINTING METHODS

Common problems

Below are inappropriate styles which are all too commonly used.

Ribbon pointing

This is seen on both old stone work that has been repointed and new stone work. It has no precedence in the past and should never be used. Each stone is circled by a ribbon of sand and cement. An attempt is made to use the pointing mortar to make each joint the same width. To achieve this, wider joints are camouflaged to look narrower, while thin joints are made to look wider. Some joints are blocked out completely while 'mock' joints are applied over the

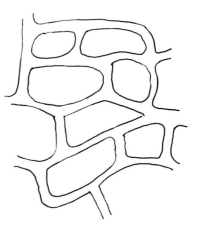

face of other stones. The use of sand and cement makes this effect permanent and waterproof. The style of pointing used is normally weather struck or strap (see below).

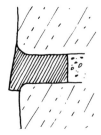

Weather struck

Common on brick and stone throughout the country. May have a place on modern brickwork but not on old brickwork or any stone work, old or new. It is normally executed in very rich mixes of cement and sand such as 1:3, it is very much a 20th century phenomenon.

Strap pointing

Wrongly referred to as tuck pointing in many parts of Ireland, strap pointing is executed in sand and cement. It is often applied to old stone buildings and walls.

Pointing trowel

The pointing trowel is an inadequate tool for pointing stone work:

- It cannot compress mortar to deep levels.
- It hangs mortar on the top arris of stones, leaving spaces behind.
- It tends to create weather struck type finishes.

POINTING ASHLAR WORK

Ashlar stone is the finest of cut stone, laid with small joints as small as 3mm in granite, limestone and sandstone. In boundary walling, where it is rare, it is generally found at the entrance to country estates and the designs are usually classical or Gothic and often includes grand piers and curved walling. Ashlar work is easier to date than most rubble walling, being 18th and mainly 19th century.

Ashlar walls are backed with rubble stone and the ashlar is merely a facade. They may have iron dogs and other fixings set in lead within their joints. When these rust they expand, this is observable on the face by spalling and cracking of the ashlar. The repair of this problem is again a job for the experts.

Any repairs to ashlar work should be made only after consideration and expert advice. Many poor attempts at pointing such work are to be seen where the pointing extends each side of the joint, or the original joints are widened using cutting blades such as angle grinders and causing irreparable damage.

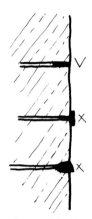

Acceptable (v) and unacceptable (x) styles of ashlar

Follow the procedure for raking out, brushing and dampening, and pointing outlined above. However, keep in mind the following:

- To conduct this work successfully, specialist trained operatives are necessary.

- Grit blasting as a means of cleaning is to be condemned as it removes surface stone with the loss of arris definition, detail and the tool marks which not only give life to the work but also represent the effort and skill of stonemasons and stonecutters in the past. The surface area of the stone as a result of grit blasting is also greatly increased and will become dirty far quicker than before. More appropriate methods such as water and poultice treatment are available from reputable stone conservationists.

- Raking-out tools need to be small and should include implements such as hack saw blades set in handles.

- A jet of compressed air can be used to clear out dust in the fine joints (wear safety equipment) and then the work can be dampened.

- Finally, the work is pointed using syringes and miniature flat bars for application and compression. Special precautions such as the use of masking tape are taken to prevent staining of the stone on either side of the mortar joint.

- After pointing, protect your work from the elements.

TOOLS FOR POINTING

Plugging chisel:

The plugging chisel is a basic tool, readily available and adequate for raking out in general work. Other specialist raking-out tools are available 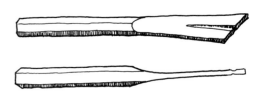 for conservation work. Raking-out tools should not create a wedging effect when used or stones will become loosened and spalled – an ordinary chisel should never be used.

Brush:

After raking out, the work is brushed clean with the head of a sweeping brush or similar.

Deck brush:

A natural stiff-fibre brush for creating a tamped surface finish on hardened mortar. It exposes the surface aggregate in the mortar so that it matches the original work. A narrow width brush is best as it will not stain the stone either side of the joint.

Hawk:

A pointing hawk is shown here with a sheet-metal top that projects on one side only by around 6mm. This pointing hawk is useful on stone with a reasonably flat plane surface and on t

Lump hammer:

1kg in weight, for use with the plugging chisel.

Flat jointing bar:

A variety of widths are required for rubble work, from 3mm to 50mm.

DOs AND DON'Ts

Do use lime mortars.

Do use coarse aggregates similar to the existing on rubble work.

Do insert stone pinnings into larger joints on rubble work.

Do finish with flush or slightly recessed joints as appropriate.

Do rake out using hand tools only.

Do pre-wet before pointing.

Do place lime mortars under compression with flat bars.

Do protect work from sun, rain and frost.

Do use hydraulic mortars where necessary such as below ground level, in very wet conditions, wall tops, etc.

Do point selectively not globally.

Do remember that pointing mortars are sacrificial and should fail before the stone.

Don't rake out perfectly good lime mortars.

Don't use electric angle grinders or petrol-driven cutting saws.

Don't use pointing styles such as weather struck and strap.

Don't use sand and cement mortars.

CHAPTER 22

WET DASHING

Wet dashing is very common on stone walls and buildings throughout Ireland. It is the traditional method of 'throwing on' a lime and aggregate wet mix to a wall in one or more coats. In Scotland it is called harling and, in parts of England, roughcast. Often only the remains of these dashes can now be seen on the wall surface in patches that are barely discernible and easy to confuse with the lime mortar used to lay the stone of the wall because their colour and character mimic the often underlying limestone structure.

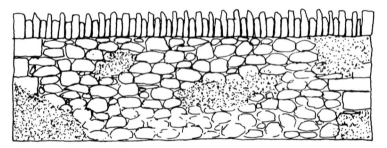

A stone wall displaying the remnants of wet dashing – it can be difficult to see the remains of wet dash on limestone walls.

After the process of weathering occurs these wet dashes appear quite flat and the colour and shape of the small stones in the aggregate are visible. As a result fresh repair work can stand out until it weathers to blend with the original work.

It seems that wet dashing was often applied as soon as a wall was completed which would imply that the stone face was never meant to be seen. This type of dashing was applied in anything from one to three coats and the same mortar source was often used to both build and dash the wall. Old dashes contain a finer aggregate than modern wet dashes.

Traditional lime dashes have a limited lifespan and should be seen as sacrificial, needing to be replaced every so often – although this could be a period of many lifetimes. Many lime dashes are now over a hundred years old and need to be repaired. It may come as a shock to some people to find that the appropriate repair procedure is not to point a wall and leave the stone exposed, but to point it and then dash it. During the raking-out procedure prior to pointing (see Chapter 21: Pointing), these dashes are often hacked off and not replaced. Behind them are hidden small irregular-shaped stones built to a reasonably flat plane surface. To point these walls is never satisfactory as with so many small stones the pointing will take up too large an overall surface area.

All of the issues discussed here apply equally to buildings. A wet dash with lime will protect a building from the weather, while at the same time allowing it to dry out quickly when wet. Note: Cement-based dashes should not be used.

HOW TO WET DASH

■ Carry out any necessary repairs, pointing and corefilling before dashing. Illustrated is a cross-section through a wall, showing pointing that has taken place and the re-insertion of pinning stones. To daub out or to fill a hollow spot on a wall small flat stones or pieces of brick or terracotta tile are used laid flat on the wall surface in lime and aggregate.

■ Mix your wet dash using lime putty with aggregate at about 1:3 and water added to produce a fairly wet mix.

- Pre-wet the wall and let it dry until just damp.
- Flick the wet dash from the trowel using a wrist action – it should make a smacking noise when it strikes the wall.
- Cover the wall lightly and evenly without leaving a pattern.

A dashing or harling trowel is used to apply the dashing.

- Do not use base coats such as scud and scratch, although deep recesses may be filled out in stages prior to dashing.
- Protect the freshly finished dash from the weather.
- If more than one coat is required, press back earlier coats when stiff with a wood float. The final coat should be left as is.
- Light coats carbonate quickly, heavy coats may be slow to harden and carbonate.
- If desired apply a number of thin coats of lime wash. The lime wash will penetrate the dashing and, through carbonation, will act to bind the dash to the wall.
- Further lime washes should be applied occasionally. The result will be a startlingly white colour initially, but this will fade gradually. Natural earth pigments may be added to give colour.

DOs AND DON'Ts

Do replace lime-based wet dashes on walls and buildings.

Do match the aggregate size, shape and colour if possible.

Do point and **daub** out as appropriate prior to wet dashing.

Do lime wash wet dashes and use earth pigments as appropriate.

Don't scud and **scratch** walls before wet dashing as in modern work.

Don't use sand and cement mixes for wet dashing.

Don't hack off old lime-based wet dashes and leave stone work exposed

High Garden Walls

High garden walls are to be seen in the grounds of many of the larger 18th and 19th century houses in Ireland. They were often built in brick and can be 3.6m (12ft) high and were used to grow fruit, etc., for the table. Brick is an excellent conductor of heat and plants ripened more quickly and earlier in the season by being grown against these walls.

Close examination of these garden walls usually reveals a myriad of holes in the lime mortar joints from the iron nails used to hold wires, etc., to train the plants. I have heard that flues were sometimes constructed in the walls, so that heat from a fire could be used to accelerate growth and keep away frost. The fruits could be quite exotic and unusual for the Irish climate. With ice collected from an ice house (usually buried underground with ice collected from a lake or pond in winter), fruit could be preserved and served cold.

High garden wall at Larch Hill, County Meath: These walls were common in large houses of the 18th/19th century and were used for growing fruit.

CHAPTER 23

STRUCTURAL PROBLEMS

Remember old walls are rarely perfect – they may lean, dip and be out of line and this may not be a cause for concern as long as it is within an acceptable limit. A certain number of holes are also acceptable and may provide homes for field mice, birds and bats. Many plants thrive on old walls and unless they are creating problems should be left alone.

The following are a number of simple structural problems encountered in free-standing boundary walls and how to treat them:

Problem: Either insufficient or no through and/or bond stones in the wall so that it becomes distorted, possibly S-shaped, particularly if it is tall or if an overhead weight has been applied.

Solution: It is generally quite impractical to introduce through or bond stones into a wall already built. In some cases a simple solution would be to insert threaded stainless steel bars with a plate on each end and nuts for tightening.

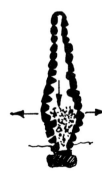

Problem: There are no through or bond stones so the core slips causing bulging in the bottom half of the wall.

Solution: If severe the only remedy may be to dismantle and rebuild the section of the wall that has been affected.

Problem: There are no through or bond stones and face work becomes dislodged.

Solution: Rebuild the collapsed section, laying stones with their length into the wall. Stainless steel tie rods could also be used as explained above.

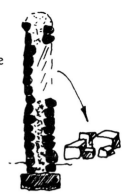

Problem: The wall is leaning because of soil pressure. The wall might not originally have been built to act as a retaining wall but, as often occurs with the boundary walls around graveyards, the soil may have gradually built up on one side.

Solution: If the wall is leaning severely and in danger of collapse it may have to be demolished and rebuilt. This presents a problem if the soil behind is part of a graveyard or an archaeological site as in either case permission is required and you may not be allowed to disturb the built-up soil. In some cases a new wall may have to be built forward of the original line.

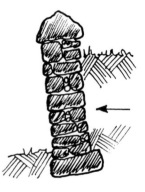

Problem: Distortion of the wall because of strong prevailing winds or soil subsidence.

Solution: In cases where the whole wall is leaning it can be jacked back into position by excavating partly under the foundations from the back and then inserting concrete from the front of the foundations. Great care must be taken to ensure that the wall is now structurally safe and will not suddenly collapse. In other cases, depending on the severity of the lean, the wall or a section of the wall may have to be rebuilt.

Problem: Differential movement of the wall because of soil subsidence.

Solution: Both lime mortared walls and dry stone walls can take a fair amount of differential movement so if the wall is reasonably sound simply leave it alone. Old walls can look attractive when they display a certain amount of distortion and twisting – it reflects their age and endurance. If severe, a limited amount of rebuild and repair may be necessary.

Problem: Wash from passing traffic causes loss of mortar in joints and individual stones subsequently become dislodged.

Solution: Rebuild washed out sections using a hydraulic mortar. If the wall leans then, depending on its width and overall structural integrity, it may be appropriate to leave it alone, attempt jacking it back into position or rebuilding.

Problem: Plant growth in wall tops and wall faces causes the wall to burst.

Solution: Remove plant, spray the wall with a biocide following instructions carefully and repair.

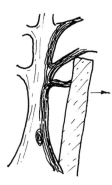

Problem: The tree is growing too close to the wall, causing it to lean. However, because trees are the habitat of many birds, insects and plants, they should be removed only as a last resort.

Solution: Consider rebuilding the section of the wall affected, only this time bridging over the tree roots. Alternatively build the wall forward of its original line. At all costs try not to interfere with the tree.

CHAPTER 24

BAD HABITS IN MODERN STONE WALLING

In recent years many miles of stone walling have been built that do not follow any traditional principles. These walls sit uncomfortably in the landscape and have little relationship to what was built in the past.

In modern stone walling the stone is often used non-structurally. Concrete cores, walls and blocks do the real work, the stone is merely a facade. Such walls are built with access to modern techniques and skills only and they reflect this in every way. Modern walls are usually accompanied by face bedding so that the walls look like vertical crazy paving. The copings consist of semi-circular shapes constructed of small stones held together with sand and cement or concrete. Sometimes there is no attempt at a coping and the walls are simply finished flat on top. To save materials most of these walls are relatively thin.

Face-bedded limestone in a modern wall looks like vertical crazy paving.

A stone wall, face-bedded and finished with a barrel-shaped coping of small stones set in sand and cement (elevation, left, and cross-section, right).

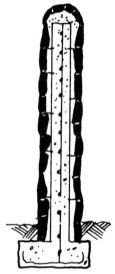

A cross-section through a modern wall showing face-bedded stone (left face), backed by concrete blocks with wall ties or expanded metal between the two and the centre filled with concrete.

A cross-section through a modern wall with face-bedded stone both sides of a reinforced concrete wall which is constructed in advance.

Similar to previous but the centre is filled with concrete in stages after the stone has been put in place. The wall must be braced to stop it collapsing while the concrete sets. The alternative is to erect shuttering in advance, lay the stone tight to the inside of the shuttering and then pour the concrete.

CHAPTER 25

INAPPROPRIATE REPAIRS

Repairs to old stone walls should be done sympathetically using the appropriate materials and skills. Nearly all problems with repairs occur because modern materials, skills and practice are being used on older work without adjustment.

The key to carrying out appropriate repairs on mortared walls is to use lime. The restraints of working with lime of necessity force a sympathetic philosophy of repair – such as the laying of stone on its natural bed. For more on lime mortars and pointing see Chapter 20: Lime Mortars and Chapter 21: Pointing.

Where holes, gaps, missing tops or partial collapse has occurred it should be repaired using similar materials and in a similar style to the original structure. Sometimes you will find the stone for this work close to hand – at the foot of the wall covered by vegetation, for example. But try not to highlight your repair work by using different materials or using traditional materials in a non-traditional way such as face-bedding stone.

Using cement will allow you more freedom than working with lime. You can, for example, lay relatively thin stone on edge to considerable heights. However, cement introduces rigidity into flexible structures so that cracking will occur. It also has an impermeable surface when applied as a render, mortar or pointing mortar so that walls are slower to dry out, and drying out can only occur through the face of the stone.

On the following page are incorrect – and all too common – methods of repairing stone walls.

A repair to a stone wall using concrete blocks.

Shuttered concrete repairs – some of these repairs are quite old and reflect the beginning of the disappearance of traditional skills in wall building and the emergence of different technologies this century.

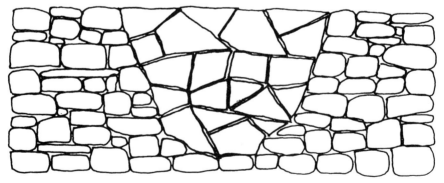

Face-bedded stone used in a repair that was not used in the original wall (it is also inappropriate to lay any stone in a style that marks it out from the rest of the wall).

CHAPTER 26

QUANTITIES, WEIGHTS AND MEASURES

The following are a number of weights and measures that are useful in calculating the quantities of material necessary for building dry and mortared walls.

per m³)	Approx Weights (tonne
Irish limestone and granite	2.70
Coarse sand	1.60
Putty lime	1.25
Cement	1.44
Water*	1.00
Lime (hydrated & bagged)	0.72

*1m³ water = 1,000 litres

STONE QUANTITIES

For 450mm wall

Quantity of stone required to build 1m² of rubble stone wall, 450mm thick
= 1m x 1m x 0.45m x 2.7 tonne = 1.2 tonne
(No allowance or deduction is made for volume taken up by loose stone, this is calculated as waste.)

For 600mm wall

To build 1m² of rubble stone wall, 600m thick
= 1m x 1m x 0.6m x 2.7 tonne = 1.6 tonne

MORTAR QUANTITIES

For 450mm wall

The number of square metres of 450mm thick rubble stone wall that can be built with 1m³ of mortar, allowing 25% of the volume of the wall as mortar (varies according to stone size and shape)
= 1/0.45 x 4/1 = 9m² per 1m³

For 600mm wall

The number of square metres of 600mm thick rubble stone that can be built with 1m³ of mortar, allowing 25% of volume of wall as mortar

$= 1/0.6 \times 4/1 = 7m^2$ per 1m³

YIELD

The volume of air space in sand is about 33%. When a binder such as lime is mixed with sand it replaces the air content of the sand without causing the sand to bulk. So 3m³ of sand with 1m³ of lime, gives a total of 4m³ of materials in all, but actually produces only 3m³. This phenomena is called yield. The same occurs when using a cement, lime, sand mix. For instance, to price a 1:2:9 mix we take the cost of 1m³ of cement, 2m³ of lime and 9m³ of sand and add these together to give the cost of 9m³ of mortar.

SUPPLIERS AND SERVICES

Company: Cornerstone (The Irish Centre for Architectural Conservation and Training)
Larch Hill
Kilcock
County Kildare
Contact: Patrick McAfee
Tel: 01 628 4518   Fax: 01 628 4538
Offers training workshops in lime, stone, etc., consultancy and products related to architectural conservation.

Company: Clogrennane Lime Ltd
Clogrennane
County Carlow
Contact: Leo Grogan/Larry Byrne
Tel: 0503 31811
Suppliers of lime putty and hydrated lime

Irish Limestone Producers (dimension limestone)

Company: Kilkenny Limestone Ltd
Quarry: Kellymount Quarry
Production: Kellymount Quarry, Paulstown, County Kilkenny
Contact: Michael Meaney
Tel: 0503 26191   Fax: 0503 26276

Company: Feely & Sons Ltd, Boyle, County Roscommon
Contact: Barry Feely
Tel: 079 62066   Fax: 079 62894

Company: James Murphy & Sons Ltd
Murphystown Road
Sandyford
Dublin 18
Quarry: Lecarrow Quarry, Lecarrow, County Roscommon
Contact: Jim Murphy
Tel: 01 295 6006   Fax: 01 295 3694

Company: McKeon Stone Ltd
Brockley Park
Stradbally
County Laois
Contacts: James McKeon/Yvonne McKeon/ David Bambrick
Tel: 0502 25151   Fax: 0502 25301
Quarry: Three Castles Quarry
Three Castles, County Kilkenny

Company: Stone Developments Ltd
Ballybrew Plant
Enniskerry
County Wicklow
Quarry: James Walshe Quarry
Old Leighlin
County Carlow
Contact: Philip Meaney
Tel: 0503 21227   Fax: 0503 21607

Limestone Rubble (calp)

Company: Roadstone Ltd
Belgard Quarry
Tallaght
Dublin 24
Tel: 01 452 5555

Irish Granite

Company: James Murphy & Sons Ltd (Dimension and rubble Granite)
Murphystown Road
Sandyford
Dublin 18
Contact: Jim Murphy
Tel: 01 295 6006   Fax: 01 295 3694

GLOSSARY

Aggregate: Naturally occurring or machine crushed stone, graded from coarse to fine particles. Good aggregate for mortars to build walls should be coarse, sharp, clean and well-graded. On average a 33% void or air space should exist in the aggregate.

Alumina: A compound of oxygen and aluminium and one of the constituents of clay.

Arris: The line or edge on which two surfaces, forming an exterior angle, meet each other.

Ashlar: Finely cut stone laid with very fine joints. (*Clocha cóirithe* in Irish.)

Bank: A bench, sometimes of stone, used to support another stone for cutting.

Basalt: An igneous stone of volcanic origin.

Batter: An inward inclination of the exterior face of a wall.

Bed: A horizontal layer of mortar; the top and bottom surfaces of a block of stone; a plane of stratification in a sedimentary stone.

Boasting: Wide chisel marks on the face of a stone.

Bond: The arrangement of stones in a wall to give structural stability and to control placement of vertical joints.

Bond stone: A stone which travels a considerable distance from one face of the wall to the other. In normal work they are at least two-thirds the width of the wall.

Boning: The sighting and paralleling of one margin of a stone with the other to take a face of a stone out of twist.

Caen stone: An oolitic limestone from Normandy, France.

Calcareous: Containing lime.

Calcination: Converting limestone (calcium carbonate) to lump lime (calcium oxide) by heat as in a kiln.

Clachan: A small group of stone houses built close together, without church or shops, and often lived in by the same extended family.

Cloch: Irish term for a stone.

Coping: The capping or covering to a wall which sheds rainwater. Prevents small stones from becoming dislodged and provides a deterrent to animals, and sometimes humans, from attempting to scale a wall.

Corbel: A stone is built into and projects from a wall to support a weight.

Corbelling: The art of placing one stone on another so that they oversail to enclose a root space or similar.

Corefill: The combination of stone and lime mortar used to fill the area between two faces of a wall.

Corefilling: The replacement of lime mortars in the heart of walls using low pressure pumping methods.

Cramp: In the past, an iron appliance used to connect two stones which was often set in lead.

Course: A layer of stones laid to a set height.

Cyclopean: Large blocks of stone used in many cultures in the past. Often face bedded and quite thin, probably used to impress.

Daub: To fill out a hollow spot on the face of a wall using mortar (to daub out).

Dimension stone: In Ireland the name given to limestone and granite taken from a quarry which can produce stone of large size and quality for accurate cutting to size.

Ditch: In Ireland a raised bank of earth or even a dry stone wall but also a cutting such as an open drain.

Dog: An iron dog – see **cramp**.

Double stone wall: A wall having two faces tied across with through stones and bond stones, the centre being hearted with small stones.

Draft: A margin worked with a chisel on the edge of a stone leaving a distinctive pattern of lines which in Ireland is called square chiselling.

Dressing: To finish a face of a stone to a particular pattern using an appropriate tool.

Dropped square: A square used for measuring depth from a face, it has one adjustable arm.

Dundry stone: An oolitic limestone imported into Ireland from the 12th century and quarried in the past at Dundry near Bristol in England.

Efflorescence: The crystallisation of salts on the face of a wall.

Face stone: A stone whose face is visible on the face of the wall.

***Feidín* wall:** A type of wall seen mostly in east County Galway and on the Aran islands off the coast of Galway, which has small stones at its base and larger stones on top.

Feldspar: A mineral consisting of silica and alumina with potash, soda or lime and an important constituent of granite. Clay is produced by the decay of feldspar.

Fence: Stone walls, raised banks and hedges are sometimes referred to as fences in Ireland.

Finial: The top piece, sometimes decorated, of a gable or spire.

Gap: Part of a dry stone wall in Ireland that can be knocked down and easily built up again to allow animals in or out of a field.

Gneiss: Metamorphosed granite with segregated layers of quartz, feldspar and mica.

Goban Saor: Legendary Irish craftsman.

Granite: An igneous stone comprising of quartz, feldspar, mica and amphibole. (*Cloch eibhir* in Irish.)

Harling: See **wet dashing**.

Hearting: The filling of the heart of a wall with stones and with lime mortar if a mortared wall. (*Cloch feidín* in Irish.)

Hot lime mix: The making of mortar by mixing lump lime (CaO) and aggregate together with water therefore combining the slaking and mixing process resulting in hot lime mortars. These were usually allowed to sour out over a period of time before using. This was by far the most common method of making lime mortars for stone masonry. Sometimes mortars were used while hot to kill frost, or in bridge construction because they were believed to result in a harder set.

Hydrated lime: Lump lime (calcium oxide) slaked and ground to a powder, often called bagged lime.

Hydraulic lime: A lime originating from a limestone, containing minerals such as alumina and silica, which give hydraulic lime the ability to set or partly set in water or damp conditions.

Limestone: A sedimentary stone laid down in sea water. Its main mineral is calcite. (*Cloch aoile* in Irish.)

Medieval: Commonly applied to the period from the Anglo-Norman invasion of 1169 to the end of the 15th century. But also confusingly applied to buildings from the 5th to 15th century.

Metamorphic stone: Igneous or sedimentary stone changed from its original form through heat and/or pressure.

Mica: A mineral with thin flexible laminae having a shining lustre. Found in granite and many sedimentary stones.

Naturally bedded: A stone lying on its flat horizontal sedimentary beds as it was originally laid down as sediment. These beds may be distorted by underground pressures which move them from a horizontal to, sometimes, a vertical position in quarries, but they should still be laid horizontally.

Non-hydraulic lime: A lime that sets or hardens through the re-introduction of carbon dioxide. The process is called carbonation.

Nose bleed: A nose bleed refers to working at the level of the soles of one's feet or even lower if excessive bending is required.

Oolitic limestone: Limestone having egg-shaped grains that form a structure similar to fish eggs. Portland stone is an example.

Out of twist: A surface taken out of winding to a true flat plane by boning and cutting.

Pig in the wall: When the string line at one end of a wall is at a different course height than it is at the other end, so that work is laid off parallel with the previous work.

Pinning: A small stone, usually flat in shape, inserted into a mortar joint while the mortar is still soft to reduce the area of mortar exposed to the weather and assist carbonation. Also built in during the building process to balance large stones or to strike a level line.

Plugs and feathers: Used since Roman times to split stones. A series of holes is drilled (originally by hand using a bullnose jumper chisel). Two metal feathers are inserted in each hole, followed by the plugs which are struck until the stone splits.

Point: A tool forged/sharpened to a point and used for removing waste stone.

Profile: Any edge such as a timber plank, specially made wooden frame, etc., from which a line can be pulled to set out and build a wall.

Punch: Does similar work to a point (see **punch**) but with a 6mm wide cutting end. In Ireland, sometimes referred to as a shifting punch.

Quartz: A crystalline form of silicon dioxide.

Quartzite: A sandstone metamorphosed.

Quicklime: Also called lump lime, calcium oxide, quicklime and CaO. Produced by burning limestone to around 900°C and driving off carbon dioxide.

Quoin stone: A term applied to cornerstones which are used at the ends of walls, at openings, etc. (*Cloch choirnéil chúinne* in Irish.)

Retaining wall: A wall designed to resist the lateral pressure of earth.

Rundale: A system of open farming widely practised in Ireland in the past. The Rundale system was usually associated with *clachans* which were small groups of houses occupied by members of the same extended family. Strips of land here and there were farmed by individuals in this group while other areas such as a bog or a field were held in common. Rundale was a complicated system because inheritance would subdivide a family's land further and break it down into still smaller strips. Just before the mid-19th century there were 2 million acres still worked under the Rundale system.

Sandstone: A sedimentary stone laid down in water or dry on land. Its grain derived from other rocks and cemented together with mud, carbonate, silica or iron. (*Cloch ghainimh* in Irish.)

Scalprachai: Irish term for a fissure in limestone on the Aran islands.

Single stone wall: A stone wall only one stone in thickness across its width.

Skew perp: A skew perpendicular joint or a vertical joint off plumb or off perpendicular with a bed.

Slaking: Introducing lump lime (also called quick lime, calcium oxide and CaO) to water resulting in a thermic reaction, followed by cooling and the production of putty lime.

Souring out: An Irish term for leaving mixed lime and aggregate in a wet condition for a period of time.

Straight edge: A flat steel bar or length of timber with a straight edge.

Template: A pattern cut in plastic, zinc or plywood to mark out stones for cutting.

Through stone: A stone which extends from one face of the wall to another thereby holding both faces together.

Wet dashing: Traditional Irish method of 'throwing on' a lime and aggregate wet mix to a wall in one or more coats. In Scotland it is called harling and, in parts of England, roughcast.

BIBLIOGRAPHY

Architectural Conservation, an Irish Viewpoint, The Architectural Association of Ireland, 1975.

Ashurst, John and Nicola, *Practical Building Conservation*, vols. 1,2 & 3, English Heritage Technical Handbook, Gower Technical Press 1988.

Carson, W. H. *The Dam Builders, The Story behind The Men who Built the Silent Valley Reservoir*, Mourne Observer Press 1981.

Bell, Feely & Meaney, *Irish Blue Limestone, Property & Applications*, An Bord Trachtala & AMSA 1995.

Bord Fáilte, *The National Monuments of Ireland*, 1964.

British Standard Code of Practice, *Cleaning and Surface Repair of Buildings*, British Standards Institution.

Building Limes Forum, *The Journal of, 'Lime News'*.

Department of Environment, *Conservation Guidelines*, 1996.

Conry, Michael, *Culm Crushers and Grinding Stones in the Barrow Valley and Castlecomer Plateau*, Conmore Press 1990.

Costello, Peter, *Dublin Churches*, Gill & Macmillan 1989.

Craig, Maurice, *The Architecture of Ireland from the Earliest Times to 1880*, B.T. Batsford Ltd. /Easons & Son 1983.

Craig, Maurice, *Dublin 1660-1860* Allen Figgis 1980.

Danaher, Kevin, *Ireland's Traditional Houses*, Bord Fáilte.

Dublin Heritage Group, *The Vernacular Buildings of East Fingal*, 1993.

English Heritage, *'Making the Point'*, brochure.

Evans, E. Estyn, *Irish Heritage*, W. Tempest, Dundalgan Press 1977.

Evans, E. Estyn, *Irish Folkways*, Routledge 1989.

Flegg, Aubrey, *Course Notes, A Geological Background to Irish Building Stones*, Geological Survey of Ireland.

Geological Survey of Ireland, *'Down to Earth'* exhibition brochure 1986.

Historic Scotland, *Preparation and Use of Lime Mortars*, Technical Advice Note, HMSO, Edinburgh 1995.

Harbison, Peter, *Guide to the National Monuments of Ireland*, Gill & Macmillan 1970.

Howe, J. Allen, *Stones and Quarries*, Pitman & Sons.

Howe, Malverd A., *Masonry*, John Wiley & Sons Inc. 1915.

Howley, James, *Follies and Garden Buildings of Ireland*, Yale Press.

Ireson, A. S., *Masonry Conservation and Repair*, Attic Books 1987.

Jackson, Patrick Wyse, *The Building Stones of Dublin*, Country House 1993.

James, John, Chartres, *The Masons who Built a Legend*, Routledge, Kegan & Paul 1985.

Johnson, David Newman, *The Irish Castle*, Irish Heritage Series, Easons & Son 1985.

Kearns, Kevin Corrigan, *Dublin's Vanishing Craftsmen, in Search of the Old Masters*, Appletree Press 1987.

Leask, Harold G., *Irish Castles and Castellated Houses*, Dundalgan Press, (W. Tempest) 1986.

Leask, Harold G., *Irish Churches and Monasteries*, vols. 1,2 & 3, Dundalgan Press (W. Tempest) 3rd ed 1987.

Loeber, Rolf, *Architects in Ireland, A Biographical Dictionary of 1600-1720*, John Murray Publications 1981.

McAfee, Patrick, 'Stonewalling', Conservation Guidelines, Department of the Environment 1996.

McKay, W. B., *Building Construction* vol. 2, Longmans 1958.

McMahon, Mary, *Medieval Church Sites of North Dublin, A Heritage Trail*, Office of Public Works 1991.

McDermott, Matthew J. & Aodhagan Brioscu, *Dublin's Architectural Development 1800-1925*, Tulcamac 1988.

Mitchell, Charles F., *Building Construction and Drawing* 12th ed 1934.

Mitchell, Frank & Ryan, Michael, *Reading the Irish Landscape*, Townhouse 1997.

Murphy, Seamus, *Stone Mad*, Routledge & Kegan Paul, 2nd ed 1986.

Nevill, W. E. *Geology of Ireland*, Allen Figgis 1963.

Office of Public Works, Ireland, *The Care and Conservation of Graveyards*, 1995.

O'Kelly, Claire, *Concise Guide to Newgrange*, C O'Kelly, Ardnalee, Blackrock, County Cork 1984.

Opderbecke/Wittenberger, *Der Steinmetz*, Callwey reprint of 1912.

Pepperell, Roy, *Stonemasonry Detailing*, Attic Books 1990.

Pfeiffer, Walter & Shaffrey, Maura, *Irish Cottages*, Artus Books 1990.

Purchase, William R. *Practical Masonry*, Attic Books, reprint of 1895 ed.

Rea, J. T. *How to Estimate*, B. T. Batsford, 6th ed 1937.

Reid, Henry, *A Practical Treatise on Concrete and how to make it, with Observations on the uses of Cements, Limes and Mortars*, R. & F. Spon 1873.

Ryan, Nicholas M., *Sparkling Granite, The Story of the Granite Working Peoples of the Three Rock Region of County Dublin*, Stone Publishing 1992.

Smith, David Shaw, *Ireland's Traditional Crafts*, Thames and Hudson, 2nd ed 1986.

Somers, Clarke & R. Englebach, *Ancient Egyptian Construction and Architecture*, Dover Publications, republished 1990.

Stalley, Roger, *The Cistercian Monasteries of Ireland, an account of the history and architecture of the White Monks in Ireland 1142-1540*.

Steinmetz Zeichen, Franz Rzina, Bauverlag reprint 1989.

Sweetman, David, 'Dating Irish Castles', *Archaeology Ireland*, vol. 6 number 4 issue number 22, Winter 1992.

Ulster Architectural Heritage Society in association with Environment Service (Historic Monuments and Buildings) *Directory of Traditional Building Skills*.

Waterman, D. M., *Somersetshire and other Foreign Building Stones in Medieval Ireland c. 1175 - 1400*.

Warland, E. G., *Modern Practical Masonry*, Pitman & Sons 1953.

Walker, *The Building Estimating Handbook*, Frank Walker & Co. Chicago, 7th ed 1931.

Whitton, J. B. *Geology and Scenery in Ireland*, Penguin Books 1974.

INDEX

A

Allied Irish Bank, Dame Street, Dublin, 35
Anglo–Normans, 25–8
Aran Islands, *99*
ashlar work, pointing, 146–7
Ashlin, George, 34
axe makers, 13

B

bad habits, 157–8
balance, 79
Baltinglass Abbey, 23
Bank of Ireland, College Green, Dublin, 31
banker mason's hammer, 49
Bankers' Georgian, 33
basalt, 39
Bearlager na Saor, 96–8
bed bonding, 74–6
bed joints, 22
bedding stone, 61–2
Bloody Foreland, County Donegal, *98*
bond stones, 54, 75
bonding, 74–9
brick dust, 132
Bronze Age, 17, 18–20
brush, 148
bull set, 49
Burren, County Clare, 18, 39

C

Caen stone, 26, 34
calp, 40
Carrickfergus Castle, County Down, 25
castles, 25–8
Cathedral, Glendalough, County Wicklow, 20–2
Ceide Fields, County Mayo, 16–18, 100
chisel, 50
Christ Church Cathedral, Dublin, 25, 34
circular piers, 118–22
cut stone piers, 121–2
how to build, 119–20
Clonfert Cathedral, County Galway, 23
Clonmacnoise, County Offaly, 22, 23, 24
Cobh Cathedral, County Cork, 96
coping, 80–3
coping stones, 55, 60
corefill, 75–6
Cormac's Chapel, Cashel, County Tipperary, 22
corner blocks, 50
crow bar, 50
cyclopean stone work, 20–2

D

daub, 123
deck brush, 148
ditches, 116–17
double dry stone retaining walls, 110–11
double stone walls, 103–4
drafted margins, 28
dry stone walling, 99–108
how to build, 105–8

Dun Aonghusa, Inishmore, County Galway, 19–20
Dundry stone, 26, 34

E

Early Christian Period, 20–2
earthquakes, 70
edge bedding, 61
efflorescence, 112
eighteenth century, 28–31
entasis, 25

F

face bedding, 62
face bonding, 76–8
face stones, 75
feidin walls, 104–5
feldspar, 39
fifteenth century, 28–31
flare kiln, 126
flat jointing bar, 148
flint, 13
fort builders, 18–20
foundation stones, 54
foundations, 70–3
calculating extent of, 71
how to lay, 72–3
Four Courts, Dublin, 32–3

G

Gallarus Oratory, County Kerry, 8, 20
garden walls, high, 153
general rubble face stones, 55
geology, 37–41
Georgian Era, 31–3
Giant's Causeway, County Antrim, 39
Glendalough, County Wicklow, 20–2
Gothic Revival, 33–4

Gothic style, 25
Gowran Master, 26, 27
granite, 37, 39
Grianan of Aileach, County Donegal, 19

H

'ha–has', 109
harling, 150
hawk, 148
high garden walls, 153
Holy Cross Abbey, County Tipperary, 68, 69
hydraulic lime, 132
hydraulic set, 131–3

I

igneous stone, 37, 39
internal corner stones, 75

K

Kilconnel Friary, County Galway, 69
Kilcooly, County Tipperary, 69
Kilmalkedar, County Kerry, 23
kiln types, 125–6

L

language (masons), 96–8
levelling, 67
lime mortars, 123–35
advantages and disadvantages, 124
hydraulic set, 131–3
mixing, 128–31
mixing methods, 133–4
non–hydraulic lime, 126–8
quantities, 161–2
types of kiln, 125–6
limestone, 39
line and pin, 49
lump hammer, 49, 148

M

'marble', 40
masons
language, 96–8
marks, 68–1
Mayo, County, 33
Mesolithic Age, 13
metamorphic stone, 41
mica, 39
mixed feed kiln, 125
mortar analysis, 138–40
mortar boards, 60
mortar mill, 128
mortared stone retaining walls, 111–12
mortars, *see* lime mortars
Mount Sandal, County Derry, 13
Mourne Wall, County Down, *12*, 100–1

N

natural bedding, 61
Neolithic Age, 13–18
Newgrange, County Meath, 14–16
Newtown Castle, Ballyvaughan, County Clare, 30–1
nineteenth century, 33–5
non–hydraulic lime, 126–8

O

O'Connell Street, Dublin, 35
O'Malley, Grainne, 28
O'Shea brothers, 34
Oxford Museum, 34

P

Pearse, William, 34
piers, circular, 118–22
pitcher, 50
plug and feathers, 51, 88–9

plugging chisel, 148
plumb bobs, 51, 66–7
point, 50
pointing, 136–49
application, 142–5
ashlar work, 146–7
brushing and dampening, 141–2
inappropriate methods, 145–6
mortar analysis, 138–40
raking out, 140–1
tools for, 148
pointing trowel, 146
Portland cement, 132–3
pozzolana, 132
pre–mixed lime putty mortars, 133
profiles, types of, 63–5
Pugin, Edward Welby, 34
punch, 50

Q

quantities, 161–2
quarry
selection of stone, 52–6
transporting stone from, 57–8
quartz, 39
quoin stones, 54, 60, 75
cutting, 88–93

R

Rathlin Island, County Antrim, 13
repairs, inappropriate, 159–60
retaining walls, 109–17
Rockfleet Castle, County Mayo, 28
Romanesque style, 22–3, 25
Rothe House, Kilkenny, 28–30
roughcast, 150
round towers, 24–5
Royal College of Surgeons, Dublin, 32
rubble stone
cutting of, 84–7

site organisation, 60
rubble walling, 42–7
random rubble built to courses, 42–3
random rubble (uncoursed), 43–4
snecked rubble, 45
squared rubble built to courses, 44
Rundale system, 31, 33, 100

S

safety, 56
St Canice's Cathedral, County
  Kilkenny, 26–7
St Doulagh's Church, Balgriffin,
  County Dublin, 23
St MacDara's Island, County Galway,
  20, 123
St Patrick's Cathedral, Dublin, 34
Saints Augustine and John Church,
  Dublin, 34–5
sandstone, 40–1
sedimentary stone, 39–41
shale, 41
sighting irons, 51
Silent Valley Reservoir Dam, 100–1
single and double combination
  walls, 104–5
single dry stone retaining walls,
  109–10
single stone walls, 102–3
site organisation, 59–60
sixteenth century, 28–31
Skellig Michael, 20
sledgehammer, 49
spalling sledge, 49
spirit level, 66
Staigue Fort, County Kerry, 19
steel square, 50
stone
history of, 13–36
how to bed, 61–2
quantities, 161

selection of, 52–6
transport of, 57–8
stone walling
bad habits in, 157–8
straight edges, 51
strap pointing, 146
striking a point, 67
structural problems, 154–6
surface finishes, 94–5

T

talus, 30–1
templates, 68–9
Temple Benan, Inishmore, County
  Galway, 22, 123, *125*
through stones, 54, 60
bonding, 74–5
Tievebulliagh, County Armagh, 13
tomb builders, 13
tools, 48–51
for pointing, 148
tower houses, 28–31
Trim Castle, County Meath, 25,
  27–8
Trinity College, Dublin, 34
trowel, 49
tuck pointing, 146
twentieth century, 35–6

V

voussoirs, 22

W

wall builders, Neolithic, 16–18
walling hammer, 49
weather struck pointing, 146
wedges, 89
weights and measures, 161–2
wet dashing, 150–2
how to, 151–2